THE HUMANOID ECONOMY

MODAR ALAOUI

WHAT PHYSICAL AI MEANS FOR INVESTMENT, INDUSTRY, AND THE FUTURE OF WORK

THE HUMANOID ECONOMY

WILEY

Published by John Wiley & Sons, Inc., Hoboken, New Jersey.

For general information on our other products and services or for technical support, please contact our Customer Care Department within the United States at (800) 762-2974, outside the United States at (317) 572-3993 or fax (317) 572-4002.

Wiley also publishes its books in a variety of electronic formats. Some content that appears in print may not be available in electronic formats. For more information about Wiley products, visit our website at www.wiley.com.

Library of Congress Cataloging-in-Publication Data is Available:

ISBN 9781394453641 (Cloth)
ISBN 9781394453658 (ePub)
ISBN 9781394453665 (ePDF)

Cover Design: Wiley
Cover Image: Modar Alaoui
Author Photo: Courtesy of Modar Alaoui

Printed and bound by CPI Group (UK) Ltd, Croydon, CR0 4YY
C9781394453641_200826

For my daughters, May and Marwa,
for my brother Mazine and my sister Maladd,
for my parents, whose light and presence continue beyond this world,
and for TQB.

Contents

Introduction *xiii*

Chapter 1 Why Humanoids, Not Just Robots 1

Robotics' Turning Point 2
The Beginning of Understanding 3
A New Foundation 5

The World Around Us Is Built for Humans 6
Designing for Humanity 7
The Payoffs 8

Making Humanoids Fit 10

The Senses Humanoids Need 11
What Humanoids Sense 11
How Sensing Works Together 12

Human Acceptance 14
Forming Acceptance 15
Acceptance and Scale 16
Form, Familiarity, and the Physical Language of Machines 17

The Co-Evolution of Humans and Technology 22
The Loop in Practice 23
How the Loop Shapes Scale 24

Chapter 2 The Missing Ecosystem I Had to Build 27

Before the Humanoid Momentum Took Shape 28
Isolated Breakthroughs 28
The Gap Between Progress and Readiness 29

Scattered Efforts Without a System 30
The Cost of Independence 31
Why the Signals Were Misread 33

When the Missing Ecosystem Became Clear 37
A Global Convergence 38
The US Landscape: Talent Without Connection 39

The Forces That Finally Aligned 41
Technical Foundations: AI and Hardware 41
Market Conditions: Cost, Operators, and Culture 42
The Power of Alignment 43

When the Path Became Clear Enough to Act 44
From Difficulty to Inevitability 45
The Decision to Act 46
The Commitment to Build 47

Chapter 3 Building the Infrastructure for a Movement 49

Launching the Venture Capital Fund as the First Piece of Infrastructure 50
The Investment Thesis 51
Why This Path 53
Past Robotics Struggles 54

The Master Plan Behind the Ecosystem I Built 59
The Builders Who Revealed the Gaps: The First Layer of Insight 60

The Humanoids Summit: The Center of Gravity for a Global Movement 61
 Establishing the Humanoids Summit 62
 The First Edition 65
 Scaling the Summit into a Global Ecosystem Engine 68
 From Global Summit to Global Platform 70

Standards and Safety: The Structure Beneath Deployment 75
 The Missing Layer in Deployment 76
 From Standards to Deployment: The Missing Layer 77
 Deployment Governance: The HOMES Framework 78

Regional Anchoring: Bay Area Robotics Association (BARA) 80
 A Working Layer 81
 Hidden Inefficiencies 81
 No Theater 83

International Humanoids Day: A Cultural Anchor for the Field 85
 What Took Shape 85
 12/12 Symmetry 86
 Shift in Perception 86

Policy Framework: The Humanoids Act Proposal 90
 Fit to Human Environments 91
 Standards in Practice 92

Regional Nodes: How the Ecosystem Became Global 93
 Complementary Regional Strengths 93
 The Need for Cross-Regional Alignment 94

Putting the Pieces Together—Infrastructure as Strategy 97
 One System, Not Many 97
 How the Pieces Reinforce Each Other 99
 Why This Field Is Different 100

Chapter 4 The Architecture of a Humanoid 103

The Five Pillars 104
Sensing: Where Reality Pushes Back First 104
Actuation and Power: Where Physics Stops Negotiating 108
Compute and Control: Where Time Becomes the Bottleneck 110
Manipulation and Dexterity: Where Usefulness Becomes Visible 113
Balance and Locomotion: Where Stability Is Negotiated 117

Integration: Why Humanoids Are Systems, Not Products 119

Failure as the Default State 121

Safety as an Architectural Constraint 124

Reliability as an Architectural Outcome 127
Specificity in Humanoid Design 131

Generalization Is the Real Constraint 134
The Variability of Real Environments 134
Why Generalization Is Difficult 135
The Limits of Optimization 135
From Demonstrations to Deployment 136
Stability, Predictability, and Trust 136
Generalization as the Bottleneck to Scale 137

Chapter 5 Funding Human-Like Machines at Scale 139

When Capital Meets Physical AI 140
The First Collision with Physical Reality 141
How Physical Risk Compounds 141

Why Physical AI Requires a Different Risk Model 142
When Capital Misreads Progress 143
How Markets Recalibrate 144

The Economic Interface Between Humans and Humanoids 145
What the Economic Interface Actually Does 146
How Work Gets Defined and Measured 147
Accountability and Trust 148
How the Interface Shapes Progress 148

The Phase Where Most Humanoid Companies Stall 150
The Metrics That Mislead 151
How Companies Escape the Stall 151

How Capital Markets Learn 152
The Signals That Confused Early Capital 152
How Markets Learned to Read Differently 153

Chapter 6 Making Humanoids Culturally Acceptable 155

Why the Fear of Human Replacement Dominates 155
How the Story Builds 156
Why the Fear Persists 158

Humanoids Will Not Replace Humans 159
What History Actually Shows 160
What Operators Actually See 162

The Work That Remains Human 162
Judgment, Responsibility, and Coordination 163

The Moment a Humanoid Enters Human Space 165
What Presence Demands 167
How Fear Changes Form 168

When Work Is Reevaluated 169
What Exposure Reveals 170
What Remains Human 171

Culture Is the Real Constraint 172
Acceptance Is Never Settled 175
What Culture Remembers 176

Chapter 7 The Global Stakes 179

Why Humanoids Become Strategic National Assets 180
What Strategic Attention Brings 181
The Risk of Fragmentation 183

How Strategic Attention Reshapes the Ecosystem 183
Standards, Evidence, and the New Currency of Trust 184
The Cost of Closing Doors 186

Fragmentation or Coordination 186
The Cost of Divergence 187
The Narrow Window for Alignment 187

Regional Gravity Without National Stereotypes 188
Four Patterns of Regional Default 189
Navigating Without Chasing Uniformity 190

Governance Becomes Infrastructure 191
Where Governance Becomes Unavoidable 192
The Infrastructure That Emerges 193

Choosing the Shape of What Comes Next 194
Where Responsibility Becomes Operational 195
Where Leadership Emerges 195

Epilogue *Humanoids 2.0—What Comes After First Contact with the Real World* *199*

The Real-World Deployment Phase 199

The Moral Decisions and Ethical Frameworks Phase 200

The Global Transformation Phase 201

About the Author *203*

Acknowledgments *205*

Index *207*

Introduction

THE SOFT CLICK of servos, the quiet whir of moving parts, the way mechanical limbs copied human motion. Those early sounds and sights grabbed my attention. I was drawn to humanoids before I even knew the word, driven by a child's curiosity about machines that could move like people.

That fascination grew on the family TV, where a steady rotation of Japanese-produced shows, *Voltron*, the live-action *X-Or*, and *Grendizer* filled the screen. My siblings and I were glued to the action, wide-eyed with excitement, convinced these robots were real. They moved with purpose, reacted like real beings, and left me wondering: What powered that kind of thinking?

Not long after, I heard the term that framed it. I remember vividly, barely three years old, hearing the phrase "artificial intelligence" on television. In that instant, everything fused: robots that looked and moved like humans, plus intelligence that could reason. From then on, I cared less about appearances and more about how robots could learn and adapt the way we do.

Even now, when the French chorus "X-Or, le shérif—shérif de l'espace!" flashes in my mind, it brings back the raw excitement of those early questions. Technology has felt magical to me ever since, but only because it feels possible.

That early pull made me see the world differently, almost like my eyes had turned into optical lenses. I carried disposable cameras everywhere, eager to understand how machines, and eventually I, could capture, read, and interpret the world. Each developed roll felt like a tiny dataset, proof that seeing was only the first step; understanding came next.

That instinct quietly pointed me toward computer vision and human behavior understanding. I started with facial micro-expression recognition, then moved into full-body movement analysis, skills that first found a home in automotive AI, where I helped pioneer in-cabin sensing software. That early work in AI wasn't about humanoid robotics yet, but it laid the behavioral foundation I would later bring into embodied AI.

Looking back, I'm not sure if it was serendipity or a dormant subconscious motivation that led me to build a career at the intersection of AI, human behavior, and user experience. After all, it's not common for an 18-year-old to wake up one day, look in the mirror, and say, "I'm going to solve how machines can interact more safely with people." Yet that's where I started, first through automotive AI, now expanding into the broader world of embodied intelligence.

The tragic car accident that affected my family served as a catalyst, bringing into sharp focus the importance of machine safety. It turned early fascinations into a purposeful path, merging my passion for AI with a deep commitment to building technology that protects people in the real world.

More than any early obsession, that event quietly shaped my direction, instilling a lasting belief: What matters most in a machine isn't just how it moves, but how it perceives, decides, and behaves around us. It shifted my thinking. I started to focus more on the behavioral side of machine intelligence, how systems interpret context, not just execute tasks. That shift stayed with me. Over time, I began approaching technical problems with a stronger sense of purpose, not just asking what machines could do, but how their decisions might affect people, and whether those actions made sense in the broader human context they operate in.

That sense of responsibility has only deepened now that I'm a proud father of two daughters. They are my everything, and they've given new meaning to my work. The joy and responsibility of fatherhood have amplified my sense of duty. When I imagine the world they'll grow up in, I don't just think about software or interfaces; I think about the physical machines they'll encounter, trust, and rely on. The thought that my daughters might one day rely on these systems when I'm not around hits differently. That parental instinct has become a new driver in my work, not just to build, but to build responsibly.

None of this would have been possible without the quiet strength of my family behind the scenes. My brother and sister have offered unconditional support, mentally, emotionally, and even financially. Their belief in me, not just as someone with ambition, but as someone they quietly understood was wired differently, has been my anchor. They embraced me as someone who never fit neatly into a box: an autotelic spirit, a recovering perfectionist, and an odd multipotentialite. In Silicon Valley terms, they supported me like a true misfit, the kind who builds his own vision rather than joins someone else's.

That support was critical in following a path that didn't yet exist. My multilingual upbringing and behavioral training at Concordia University, then under the umbrella of integrated marketing communication, sharpened my ability to read people and interpret context with nuance, especially in cross-cultural settings. Combined with years of travel, this became second nature and later proved essential in designing systems that need to read human behavior in dynamic environments.

These memories and experiences are only fragments of a larger story, one I'm borrowing from a future memoir. But for now, they serve as a bridge from my earliest wonder with humanoid machines to the movement I've helped build around them.

Over time, that movement took shape across multiple layers. I founded ALM Ventures, launched the Humanoids Summit across the United States, Europe, and Asia, created International Humanoids Day, contributed to the drafting of US humanoid safety standards, and drafted a proposal for a US Humanoids Act to unlock federal funding and strengthen long-term US leadership in the field. Each of these efforts reflects a larger goal: Building the capital, infrastructure, policy, and cultural foundation needed to scale humanoid systems responsibly.

None of it happened by accident. The same instincts that drove me as a child—curiosity, a deep concern for how machines interact with people, and a need to understand how systems behave—still guide my work today. From a robot-obsessed kid to a Silicon Valley technologist and global contributor to the field of embodied AI, the path might look serendipitous from the outside. But for me, it was driven by pattern recognition and a growing sense of conviction. What began as a fascination became a conviction. That conviction became a mission. That mission became my life's work.

The first chapter of that work began more than a decade ago. I started building real-world AI systems before humanoids entered the spotlight. Leading early efforts in computer vision and human behavior understanding gave me a front-row seat to the rise of embodied intelligence. While others were focused on abstract models, I was working on how machines perceive and respond in physical environments, with real-world safety, motion, and intent in mind.

This early exposure helped shape an intuition for long-cycle technologies. It also built an internal sense of responsibility, knowing that every advancement in how machines behave around people carries weight. As the field grew, so did the recognition: I've had the chance to keynote global stages and speak at hundreds of conferences across the United States, Europe, and Asia, sharing a vision that bridges technical depth with practical impact.

From my base in Silicon Valley, I've collaborated with global hardware and AI leaders and traveled to the core hubs of advanced manufacturing and robotics from Tokyo and Seoul to Munich and Boston. These experiences expanded not only my technical perspective but also my understanding of the different cultural lenses through which we view human-machine interaction.

What continues to energize me most is the engineering itself. I still take deep satisfaction in solving tough systems problems with fellow engineers. I've layered technical depth onto my business background, especially since the rise of modern deep learning. That edge lets me navigate both the low-level design of intelligent systems and the strategic context required to scale them safely.

I've served as both CEO and CTO, a dual role that gave me a rare 50,000-foot view of the field while staying close to the technical details. That combination allowed me to bridge deep engineering work with broader product and business impact, a perspective that still shapes how I approach embodied AI today.

Over the years, I've added substantial technical depth to my business foundation, especially with the rise of machine learning and deep learning. That journey has led me to passionately geek out on low-level system design with fellow engineers, an achievement that feels equivalent to earning a double PhD, and one I'm genuinely proud of.

As we enter the age of embodied intelligence, the machines around us are becoming more than just tools. They're starting to perceive, decide, and act in ways that echo human behavior. Humanoids represent a convergence point in robotics, AI, hardware, and interaction design. But what's often overlooked is that their success will depend not just on how they function, but on how well they align with human needs, values, and expectations.

This book isn't about hyping robotics or speculating on far-off futures. It's about unpacking the real momentum behind humanoids today, from research breakthroughs to commercialization efforts, safety standards, industrial deployment, and beyond. Along the way, I'll share the patterns I've seen emerge and the lessons I've learned helping build the ecosystem from multiple angles: investor, convenor, founder, and technologist.

Humanoids are no longer sci-fi. They're here, and this decade will define how they scale, how they serve, and how they earn our trust. I've had the chance to help shape this movement from the inside across research, startups, investment, policy, and global convening. And in this book, I'm opening that vantage point to you, not as a futurist, but as someone who's been working at the front lines of the humanoid robotics ecosystem, seeing how the pieces fit together, and why it's taken this long to get here.

This book isn't about imagining distant futures. It's about unpacking what's happening now, what's working, what's missing, and what it will take to move forward responsibly. I'll walk you through what I've seen, what I've helped build, and what's still taking shape, from the labs to the factory floor to the capital behind it all.

If you're curious about how this all came together, and where it might lead, you're in the right place.

Let's step into it.

1

Why Humanoids, Not Just Robots

My earliest memories are from the back seat of my family's car, rocking gently with each acceleration and stop, laughing as my father played along with my requests to speed up or slow down. I couldn't see the road, but I could feel the rhythm of the machine. The way it moved with us. The way it responded.

Even then, something about that precision and control felt alive. Not exactly human, but something that hinted at it. I didn't have the words for it then, but I remember feeling emotionally connected to the machine itself, as if it were participating in our experience, not just carrying us from one place to another.

Years later, I became deeply involved in robotics. I saw machinery as an avenue for safety, not just as a vehicle for production. With each advancement in technology, it felt as though we were becoming more integrated all the time.

That feeling deepened the first time I saw a robot shaped like a person. Something about the human form, even when rendered in clunky metal or exaggerated plastic limbs, felt different. It wasn't just another machine doing a job. It was something crossing over. Something stepping, even awkwardly, into our space.

But something was incomplete. The mechanics were there, but there was no understanding. I felt the threshold, even if I couldn't name it.

It must have been the connection I'd made when I was a child, though I hadn't recognized it at the time. Just like in vehicles, the speedometer needle moved, yes, but the room inside the car also shifted. My father's hand eased or firmed on the wheel, and the sound of the engine rose and settled. The same input meant different things depending on what surrounded it.

I suddenly realized these were principles rooted in reality, not abstract as I'd once thought. That is how people live. We read intention from posture and timing long before words arrive. As I grew into the work, that early sense became a rule of thumb. A camera can count pixels on a crosswalk, but the decision to yield happens because a person angles their shoulders; because a foot hovers, then commits; because a stroller's front wheel noses toward the curb. A microphone can hear a tone, but the reason to stop is the direction of a shout and the urgency in it. Touch sensors can report force, but the right grip is set by the wobble that hints at a slip, not by a single reading. Inputs tell you what is there, and context tells you what it means.

I noticed the same lesson in the small glitches of daily life. An automatic door that opens late because a child's height falls outside its comfort zone. A sink that will not start for a person wearing dark sleeves. A parking sensor that beeps wildly at a harmless bush while missing a low concrete post. Each miss is a reminder that sensing without context is brittle: It works until the world asks for a nuance the system never learned.

That is why my work moved toward behavior, not only where people are, but what they are likely to do next. Not only how a machine moves, but how its movement will be read by the people nearby. The safest path, the calmest floor, the product that earns trust, all come from systems that treat scenes as living patterns rather than as isolated measurements.

The feeling from the back seat was right: Machines will belong in our spaces when they learn to move with our rhythm, not just react to our data.

Robotics' Turning Point

Robotics has always been defined by fixed rules and predictable routines. Robots were strong and precise, but their abilities were confined to narrow boundaries. They relied on structured conditions and were easily disrupted when anything unexpected entered their space. Even the most advanced systems could repeat complex actions, yet not understand what those actions meant. Robotics was useful, but it lacked awareness.

Just a few years ago, robots were closer to electromechanical and mechatronic tools—machines programmed to follow instructions—than to capable partners. A robot in a restaurant could carry dishes, but only along taped routes. A hotel robot could bring towels to a room, but only if no one interrupted its path. A logistics robot could move bins from point to point, yet a single cart out of place brought it to a stop.

These systems worked, but only because people adjusted their environments to accommodate them. Robots did not adapt to the world. The world adapted to robots. And adaptation required changes in infrastructure, something the world wasn't ready to commit to.

It was not until the rise of artificial intelligence, aided by advances in generative AI, foundation models, and early research on world models and spatial AI, that robotics began to reach its turning point.

These branches of AI signaled a broader shift. Intelligence was no longer limited to processing data. It could interpret context, anticipate outcomes, and adapt to situations it had never seen before.

Once that kind of intelligence met a physical machine, everything changed. Robotics moved from rigid execution to flexible understanding, and the field crossed a threshold into what we now refer to as physical AI, where intelligence is not just computed, but embodied.

The Beginning of Understanding

The change began as learning systems replaced rigid programming. Learning accelerated first. Instead of relying on countless hardcoded rules, robots learned from data, demonstrations, and experience. They improved through reinforcement learning, imitation learning, and well-designed simulation pipelines. A robot could practice thousands of variations of a task overnight, then bring that learning into the real world the next morning.

As these systems matured, vision models had matured enough to understand not just objects but the relationships between them. Audio models recognized direction and urgency rather than simple volume. Language models made it possible to give instructions in plain speech. Touch sensing became precise enough to adjust grip or recover from a slip.

This represented a new foundation for robotics. Instead of treating every frame or reading as separate, robots learned to connect the dots.

Locomotion, dexterity, and full-body coordination advanced at a pace robotics had never seen before, taking the rules given and translating them into motion. While visual and audio models progressed, robots learned to climb stairs with balance, recover from bumps, open doors, carry objects across uneven surfaces, and handle fragile items with controlled force. Movement became fluid instead of mechanical. The next step relied heavily on artificial intelligence.

As these capabilities advanced further, systems became more adaptable. Robots stopped depending on perfect conditions. They handled imperfect lighting, cluttered tables, mixed traffic, and small surprises that used to cause shutdowns.

Communication improved as well. Robots could understand speech more reliably. They recognized gestures, like pointing or waving. They responded to uncertainty in ways that felt natural for people. A server pointing to a table. A front desk agent waving a robot through. A worker telling a robot to "hang on." These small interactions became part of normal operations because the robot understood them.

The turning point had become visible: The environment no longer had to change for the robot; the robot adapted to the environment.

Robots no longer felt like machines forcing their way through a space. They behaved like systems that understood how to participate in that space. The experience shifted from managing a robot to working alongside one.

The transition first became visible in public-facing environments, where robots had to operate alongside people rather than inside controlled conditions. These settings introduced variability immediately. Movement was unpredictable, interruptions were constant, and context changed moment to moment.

Early systems were functional but rigid. They moved carefully, depended on cleared paths, and often required intervention when conditions deviated from expectations. A single obstacle could break the flow of operation. The system worked, but only when the environment behaved as expected.

As perception, learning, and adaptation improved, behavior began to shift. Systems became more responsive to their surroundings, adjusting to movement, rerouting without supervision, and handling small variations that previously caused failure. The change was not in isolated capability, but in how systems behaved under continuous change.

As these capabilities matured, the next layer emerged. Systems began to show early forms of reasoning. They moved from reacting step by step to anticipating what might happen next. Behavior became more fluid, with actions shaped not only by what was detected, but by what was likely to follow. This shift remains in progress, but it continues to expand how systems operate in real environments.

In hospitality settings, this change became visible in how robots moved through shared spaces. They adjusted to foot traffic, navigated around obstacles that were not present moments earlier, and paused or yielded when people approached. Movement became coordinated rather than scripted. The environment did not change for the robot. The robot adapted to the environment.

The same pattern appeared in restaurants and large mixed-use facilities. Systems adjusted their pace based on activity, slowed near intersections, and anticipated movement rather than reacting late. In logistics environments, timing and spacing adapted to human activity instead of following fixed schedules. Behavior became responsive to the flow of the environment rather than constrained by it.

Robots began operating in dynamic settings where people already are, not as systems that required accommodation, but as systems learning to participate.

A New Foundation

With these layers combined, robotics shifted from narrow tools to general-purpose platforms. A single system could move across environments, from hospitality to healthcare to logistics, without requiring those environments to be redesigned.

The world is built for human bodies. A robot that shares our reach, height, and movement can participate in that world with the least friction. AI gives robots the intelligence to do it.

The robotics turning point is not defined by a single breakthrough or model. It is defined by the moment intelligence became physical. Robotics gained the ability to sense, learn, move, and collaborate in environments designed for people.

The World Around Us Is Built for Humans

Everything around us was built with one set of measurements in mind. Ours. Doorways are roughly double the width of our shoulders. Tables meet us at the height of bent arms. Steps match the length of our stride. Handrails, light switches, and door handles are all within easy reach. From the curvature of a chair to the angle of a staircase, the physical world is tuned to human dimensions. The goal was simple: Make the environment usable for as many people as possible without special tools or modifications.

It also meant the entire world around us evolved to complement our movement.

Floors are flat to accommodate walking. Shelves are positioned for a typical arm span. Tools are shaped for hands with opposable thumbs. This consistency makes it easy to navigate unfamiliar places without thinking.

This becomes a problem for any robot that isn't human-shaped. Most robots until now have been physically incapable of operating in this environment. Industrial robots thrive when their surroundings are built for their purposes: flat floors, fixed workstations, predictable layouts. The real world is cluttered, irregular, and full of obstacles, and people move unpredictably.

For a robot to operate here, it needs adaptability. It must recognize objects and spaces in context, understand how they relate, and adjust its actions in real time. The problem is that specialized robots do not generalize well to the rest of the physical world. The moment the task crosses rooms, stairs, and mixed traffic, the machine becomes a project.

A robot that can move, reach, and manipulate objects like a person does not need the world rebuilt for it. It can walk through doorways, climb stairs, and use the same tools and equipment as people. The infrastructure is already there. So, that is where the next generation of robots must go.

And humanoids are designed for exactly that.

Their form is not a stylistic choice. It is an engineering solution to a problem that has been with robotics from the start. By giving robots human-like height, reach, and mobility, we give them the keys to the environment we already built.

Designing for Humanity

To see the argument clearly, the geometry must be specific.

Early humanoid prototypes struggled with ordinary shelf layouts where bins are placed between mid-thigh and low-chest height. The arms could reach the bins, but the wrist could not rotate to present items in the way workers were used to receiving them. Staff began compensating by lowering their posture or adjusting their grip, which slowed the workflow. After designers aligned shoulder height, arm length, and wrist rotation with common human reach, the handover became natural, and the process regained its normal pace.

The results proved that even small anthropometric choices change outcomes. A few inches in shoulder height decide whether a robot can see over a cart and read a door swing. A few degrees in hip rotation decide whether it can pivot in a narrow hallway without rubbing the wall. A little more dorsiflexion at the ankle decides whether it can descend stairs steadily with a load. On paper, these sound minor. In a real environment, they decide whether a team trusts the machine to move alone.

But movement becomes much more complicated. It's not just about movement. The way a humanoid interacts with tools, machinery, and furniture has to match human ergonomics. A robot arm that is too short or a grip that is too weak will fail at tasks people take for granted. The design must account for the force required to open doors, the dexterity needed to plug in cables, and the precision needed to pour liquid without spilling.

The solution is reach. Reach is three-dimensional. Most work zones put frequent touches between low-chest and mid-thigh for a standing adult. That is where parts bins live. That is where tools hang. That is where handovers happen. A humanoid with similar shoulder height and arm length can work within those parameters without redesign. If the wrist can rotate to present a tool in a familiar way, the handover is easy. If it cannot, people will compensate, and the flow will suffer.

Grip follows the same principles. Tool makers shape handles to fit a human hand, not to fit a claw. A good humanoid hand does not need five perfect fingers. It needs a stable opposition grip, a couple of fingertip force sensors, and a wrist that can change orientation without fighting torque. The proper grip is enough to hold a drill, turn a lever, and carry a tote without bruising it. When the shoulder and elbow are in human-like positions, the wrist has room to do its job.

Where the arms go, the feet must follow suit. Locomotion matches the built world, too. Most floors have small gaps and thresholds. Stairs are common. Ramps vary in slope. A wheeled platform is efficient on perfect floors, then pays a tax the moment a gap or a stair appears. A legged platform pays an energy cost to step, then recovers it through coverage. In many sites, coverage matters more than peak speed. Coverage determines whether one unit can work across rooms, between floors, through doorways, and whether the buyer avoids construction.

Though complexity is inherent in humanoid creation, they don't need to match humans completely. The goal is compatibility, not mimicry. In many cases, functional equivalence is more important than anatomical replication. A humanoid might have a hand with three fingers instead of five, or joints that rotate differently, as long as it can grasp the same objects and perform the same tasks.

The Payoffs

The economics point in the same direction. Fitting machines into human spaces preserves capital and allows work to continue.

Baseline fit means simpler evaluation. Many sites use a simple fit screen, often a single page, to judge potential tasks. These checks confirm whether the task can be attempted without altering structure or fixtures. When the screen is green, a trial moves forward with only small supporting details like labels, a staging zone, or a clear approach path, limiting physical changes to what is already human-standard.

Human geometry is not uniform across the world, but it rhymes. Handle shapes differ. Labels differ. Traffic flows left in some places and right in others. Stair proportions vary. The solution is a body that sits inside a tolerant human envelope, then carries region packs for language, signals, and a few adapters. That keeps hardware stable and keeps learning transferable.

Evolution means humanoids and humans work in tandem. A humanoid trained in one environment can move to another where human geometry is similar. Minor parameters adjust—such as aisle width, door swing, preferred approach speeds, or handover height—and the robot continues with short recalibration.

In practice, the pattern is simple. We might imagine that a mid-sized distribution center with recurring slowdowns moves totes from inbound to the sortation area. The geometry was aligned with human standards, so a humanoid could run the same route without altering the environment.

The first week brings high interventions and frequent pauses at intersections, each captured in short clips and fed back into training. By the second week, intervention counts fall and timing improves around shared lanes. By the third week, cycle-time variance settles, and the team adds a second route that requires no construction.

A month later, the same skill transfers to a neighboring building with only small adjustments to door swing and preferred passing etiquette. Approvals move quickly because incident clips were short, labeled, and trending down.

The result: The setting stayed open, the workflow remained familiar, and the expansion timeline shrank from months to weeks.

Over time and with enough consistency, a standard test suite takes shape: narrow hallway yields, door handling, mid-shelf picks, clean handovers. Benchmarks shift from novelty to reliability. Buyers compare behavior in familiar scenes, not curated videos. Results hold across quarters, not only at launch.

Consistency like this builds trust beyond the operational team. Regulators and insurers favor changes that do not disturb physical structure or emergency access. They also favor environments where incident trends show a steady decline. A site that records short incident clips, maintains a clear taxonomy, and shows falling interventions over a pilot period demonstrates control. That pattern earns approvals and better rates.

Momentum becomes the final advantage. Projects that avoid reconstruction finish. Completed projects generate evidence. Evidence builds trust. Trust accelerates the next deployment. Fit becomes the quiet engine behind progress, letting robots evolve inside the world as it is, without asking the world to reorganize around them.

How well a humanoid fits has limits, but those limits clarify its value. Some tasks require tools built for occasional use, which do not justify early adaptation. Others cross access points that cannot be navigated safely. Some rely on a unique fixture that makes sense only at one site. A mature program acknowledges these limitations.

In the end, form follows the world. The world we built follows us, and a humanoid form fits because our buildings, tools, and habits already assume it will. When robots match that assumption, integration gets cheaper and schedules get shorter. The result is a practical alignment with the geometry of daily life. That is the unlock.

Making Humanoids Fit

The built world already reflects centuries of human measurements, human routines, and human reach. When robots adapt to those standards, work continues as it is. People stay in rhythm. The proof phase moves faster because nothing upstream slows it down.

This perspective follows naturally from the logic in the previous section. Human geometry is already encoded in buildings, tools, and workflows. That shared standard has always been a constraint for robotics.

With humanoids, it becomes an advantage. The more a robot aligns with this geometry, the more locations it can enter without friction.

Any construction, even a small one, disrupts the surrounding environment. A single change to a doorway triggers permits, inspections, and temporary shutdowns. A modified aisle affects traffic, schedules, and shift flow. Teams lose momentum while workspaces wait for signoff.

Even when upgrades move quickly, they push everything else aside. When the environment remains untouched, normal operations continue, and the robot trial blends into the existing pattern.

But the economic benefits for workers are even greater. Workers keep their patterns. Supervisors maintain their schedules. Training stays short because the environment does not change. Operators who used to carry totes shift into roles focused on guidance and oversight. Morale improves because the room stays familiar, and strain drops.

The way humanoids fit into their environments determines their long-term flexibility. Fit preserves optionality. A space reshaped for a single form of automation becomes dependent on that design. If a newer platform appears, switching becomes costly. When the setting remains aligned with human standards, multiple vendors can operate in the same building with minimal adjustment. Renewal cycles reflect performance, not sunk construction.

The financial picture changes as well. In the traditional model, capital flowed toward construction: widening corridors, adding markers, or installing machine-specific infrastructure. With fit as the default, capital shifts toward service, learning, and iteration. A robot becomes a movable asset. Skill graphs travel with it. Depreciation follows the hardware rather than the building.

The Senses Humanoids Need

Humanoids can't fit into environments they can't sense. A humanoid must see, hear, and feel in ways that align with human expectations in shared environments. It must also understand its own body well enough to move steadily without constant correction. Each sense matters only when it changes a decision in time.

What Humanoids Sense

Understanding what humanoids needed to sense became the most important step in developing a functioning robot.

Vision carries most of the load. Effective systems rely on a head cluster with a wide field of view, supported by depth sensors and a small number of auxiliary cameras near the hands and ground. This layout avoids occlusion during common motions and keeps maintenance straightforward. Cameras positioned where people naturally expect eyes to be make behavior easier to interpret, and avoiding impact zones reduces wear. Cleaning becomes a quick reach rather than a task that needs tools or special access.

Depth can come from stereo, structured light, time of flight, or a blended approach tuned to the setting. The choice depends on the work. Shelf interactions benefit from stable short-range depth, while navigation in mixed traffic benefits from consistent medium-range motion cues.

Hearing offers additional context. A cluster of microphones near human ear height gives localization. Models listen for a small set of cues: a stop word, a shout, a directional call, a name, or a distinct industrial sound like a cart alarm. What matters is meaning and distance. A loud call in a clinic becomes a yield signal. The same call in a distant part of a warehouse is ambient. Clear, predictable policies help people understand how the system will react.

Touch closes the loop in the hands and load-bearing joints. Fingertip and palm sensors report grip force and slip. Torque sensing at the wrist and elbow detects binds before damage. Force sensors in the feet help maintain balance on uneven surfaces.

Proprioception ties the body map together. Joint encoders, gyros, and accelerometers track pose and motion without relying on vision. This allows a humanoid to reach for a tool clipped to a belt, climb a short step, or carry a box while glancing at signage. The controller trusts the internal map first and uses vision for confirmation.

Environmental sensing extends capability where humans have gaps. A compact thermal camera identifies heat sources that may indicate equipment stress. Gas sensors detect harmful conditions early. These additions are selective and appear only where they influence behavior rather than to expand a specification list.

How Sensing Works Together

Sensor fusion is as much about timing as it is about mathematics. Cameras, microphones, encoders, and force pads speak at different rates.

A shared clock aligns them. Stale packets drop out. Systems favor fewer signals used well over many signals used poorly. If a grasp needs to adjust within tens of milliseconds, the force signal must arrive on time, or it becomes irrelevant. It all comes down to the combination of key signals.

Latency budgets guide everything. Vision to planning to control has a target. Audio cue to yield has a target. The force spike-to-grip adjustment has a target. Missing these windows leads to hesitation, bumps, or drops. Meeting them keeps the environment calm. These numbers define how behavior feels to people nearby.

Edge compute keeps loops tight. Policies run locally. Heavy learning runs in the cloud or a data center using clips that respect privacy rules. If a connection drops, the system falls back to safe local behavior. Motion near people never depends on a network trip.

Context filters matter. A posture means one thing at close range and another at a distance. A sound means one thing in a clinic and another in a warehouse. The model adapts its expectations based on the environment and time of day. These filters are modest but make the robot behave like a steady teammate.

Placement follows practical rules. Sensors sit where people expect them, avoid self-occlusion during everyday movement, and are easy to clean or replace. This is why many strong designs cluster vision in the head, distribute tactile sensing across the palms and fingertips, and place a small number of body cameras to cover blind spots. Unconventional layouts often introduce interpretation costs or maintenance burdens that compound over time.

Calibration works best when treated as a short ritual. A brief check at the start of a shift confirms balance, joint angles, and camera color accuracy. A tap pad verifies fingertip force. A tone confirms microphone health. A clear pass or fail result builds trust. If a sensor drifts, the system schedules a pause at the next safe point and requests attention. Reliability comes from consistent self-verification, not ceremony.

Maintenance is part of sensing design. Lenses scratch, cables loosen, and tactile pads wear. A weekly wipe, monthly alignment, and quarterly deep check prevent most surprises. Spares in a small kit allow quick swaps. The console tracks hours and recommends service before failure. Predictability keeps operations calm.

Redundancy is selective. Safety-critical situations benefit from overlapping cues, such as combining a stop word with a visible palm-out gesture, or checking both distance and speed during close interactions. Redundant sensors everywhere add weight, power draw, and maintenance overhead. Effective systems keep what demonstrably changes behavior.

Testing remains practical and consistent across sites. A humanoid walks a basic route marked for stops, yields, door interactions, and handovers. Cameras must read labels at set distances. Microphones must localize a shout from a marked position. Tactile pads must register both a light touch and a firm grip.

Data handling is simple and legible. Clips are short and tagged in plain language, such as "narrow hallway yield," "door swing misread," or "grip slip on carton." Reviews focus on improvement, not blame. Fixes come from policy edits or small layout adjustments. Wins circulate so other sites can copy them.

Failure modes sit inside the design. Cameras may be blocked, audio may misfire, or a tactile pad may fail. The system detects the issue, slows, yields, or stops calmly, then resumes when safe. People remember how a machine behaves under stress, and calm failure earns trust.

The goal is to sense enough, fuse well, and act in time so that work flows. Everything else is a future step. Eventually, a humanoid stops feeling like a moving frame and becomes a reliable partner in environments designed for people. With fit and sensing established, the next question is how people receive the machine.

Human Acceptance

Human acceptance sits at the center of the humanoids' movement. It is the one signal that cannot be engineered through marketing or accelerated through raw technical progress. Acceptance forms in real space, through real exposure, in front of real people who did not plan to share ground with a machine.

Humanoids are ultimately accepted because they align with human expectations. People read posture before they read capability. They read motion before they read intention. They read rhythm before they read purpose. A machine that moves inside the same unwritten rules that guide human flow opens a path toward trust. A machine that ignores those rules never quite fits, no matter how advanced its hardware.

Forming Acceptance

Across the environments where humanoids appear today, the pattern repeats. When a humanoid enters a space, the air shifts slightly. People straighten. Their pace adjusts. Conversations pause for a brief moment. Eyes lock onto the machine before drifting back to familiar routines. This instinctive recalibration is the starting point of acceptance.

Acceptance is not earned through novelty, and novelty produces attention, not trust. Trust grows through predictability. A person watching a humanoid take its first steps inside a demo hall is not evaluating technical performance. They are evaluating whether they understand what the machine is about to do. If the robot glances toward a direction before turning, people settle. If it angles its torso slightly before stepping forward, people interpret that motion the way they interpret each other. If its pace stays steady and consistent, the crowd relaxes. These are subtle cues, but they carry weight. They create the sense that the machine lives inside the same invisible map everyone else uses.

Even in demonstrations I've curated, the strongest signal always comes from the crowd. A humanoid steps forward, the room forms a natural arc, and instinct takes over. People track the shoulder line, the way the arms settle, the steadiness of the steps, and whether the machine respects their space. The excitement is real, but it is paired with an immediate need to understand the machine's rhythm. When that rhythm feels natural, the room lifts. When it feels off, unease settles in just as fast.

This pattern of acceptance is remarkably consistent across different environments. Public exhibitions often attract visitors with little familiarity with advanced robotics. Research labs and fulfillment centers, by contrast, bring audiences who understand the technology in far greater depth. Yet even among experts, instinctive reactions come first. People watch posture, movement, and behavior before evaluating technical capability. Acceptance is earned through behavior, not form.

Early fulfillment center pilots add real-world pressure. Workers immediately watch how the machine handles turns, responds to nearby movement, and maintains stability around carts and stacked goods. A humanoid that respects pace, maintains a clear posture, and telegraphs intent becomes easier to work around. Workers return to routines faster.

Human acceptance is observed and felt. People need to *see* that the machine understands boundaries without being explicitly told. They need to *see* that it behaves consistently across minutes, not just during a choreographed moment. They need to *see* that its presence does not demand constant attention.

Human acceptance also carries a cultural layer, but it shows up lightly at this stage of the movement. Different regions have different comfort zones with public robotics. Some crowds prefer more distance. Others prefer a friendly gesture. Some environments tolerate noise. Others do not. The instinctive pause, the curiosity, the steady shift toward familiarity, and the evaluation of predictability appear almost everywhere in the same sequence. Cultural differences will matter more as deployments scale, but right now, the foundational acceptance curve looks universal.

Organic acceptance always outperforms forced acceptance. No amount of staging or scripted demonstrations can replace genuine exposure in real space. The strongest acceptance stories come from ordinary moments: a humanoid maintaining a clear posture while someone walks behind it, a humanoid slowing slightly as it passes a group during a demo, a humanoid holding a stable handover position without wavering.

These moments show that the machine respects the environment. They show that it understands the human rhythm. They show that technology can enter shared space without taking anything away from the people in it.

Acceptance and Scale

Human acceptance is critical to deployment and determines the speed of adoption. Humanoids are not entering the world as isolated tools. They are entering as physical AI systems in human-shaped form.

People expect them to understand context, read signals, and navigate without friction. These expectations are natural outcomes of the shape itself.

A humanoid that creates tension slows everything around it, from stakeholder buy-in to on-site training to regulatory comfort. A humanoid that generates calm speeds adoption, and calm is the highest signal of acceptance. Calm shows up in posture, tone, movement, and eye tracking. Calm shows up when people walk past a humanoid without adjusting their own motion. Calm shows up when the machine becomes part of the background rather than the center of attention. Calm is the point where integration begins.

Acceptance is what allows the movement to scale responsibly. It ensures that humanoids help rather than disrupt. It keeps deployments grounded in human reality rather than technical ambition. It opens the door to collaboration, trust, and real-world permanence. When acceptance holds, training shrinks, onboarding lightens, friction fades, and the machine becomes part of the environment rather than a guest in it.

The world is entering a phase where AI is no longer confined to screens. It is stepping into physical form. That shift changes everything: how people think about AI, how industries adopt new systems, how spaces are designed, and how workflows evolve. None of that acceleration can happen without human acceptance at the center.

Form, Familiarity, and the Physical Language of Machines

The role of acceptance brings attention back to a deeper structural reason humanoids matter. Technical capability alone rarely determines which technologies succeed in human environments. Compatibility with existing systems often matters just as much. In the case of humanoids, that compatibility begins with form. While robotics has produced many highly capable machines over the decades, most of them were designed for environments that were deliberately engineered around the machine. Humanoids approach the problem from the opposite direction by aligning themselves with environments that were already engineered around the human body.

Human civilization has quietly standardized itself around a set of physical assumptions. The height of door handles, the spacing of shelves, the rise of stairs, and the placement of switches and controls all reflect human reach and posture. Workflows across industries evolved under the same assumptions. Workers pass tools to each other at a certain height. Objects are stored where the human arm can reach comfortably. Hallways allow two people to pass without friction. These patterns exist everywhere from factories and hospitals to hotels and office buildings.

When a machine shares those proportions, it inherits immediate access to the world that was designed for people. This does not mean a humanoid must perfectly replicate human anatomy, but it does mean that approximate human scale creates enormous advantages. A machine that can reach the same shelves, grasp the same handles, and step over the same thresholds can enter environments without requiring those environments to be rebuilt.

For most of robotics history the opposite approach dominated. Engineers placed robots in carefully controlled spaces where everything was predictable. Industrial arms operated behind safety cages. Warehouse robots followed taped routes across floors designed specifically for them. Service robots moved along predefined corridors that staff had been trained to keep clear. These machines performed their roles extremely well within those boundaries, but the boundaries themselves were part of the design.

Humanoids remove much of that requirement. Instead of modifying the environment to match the robot, the robot is shaped to match the environment.

This shift may appear subtle at first, yet it has far-reaching consequences for deployment. Every structural change to a building introduces friction. Construction triggers permits, inspections, and operational interruptions. Even small adjustments to workspaces require teams to pause routines that may have taken years to refine. When machines enter spaces without requiring those changes, adoption becomes significantly easier.

The impact becomes clear when observing how people respond to machines in real environments. During demonstrations and pilot deployments, the technical performance of a robot certainly matters. Engineers pay attention to balance, manipulation, and navigation. Yet the immediate reaction from most observers is rarely about those details. Instead, people focus on whether the machine appears to understand the space it occupies.

When a robot approaches a doorway and aligns its body naturally with the opening, the movement feels intuitive. When it reaches toward a handle in a way that resembles human posture, observers immediately understand the intent. When it pauses slightly to allow someone to pass in a hallway, the behavior reads as respectful of shared space. None of these actions require complex explanation because people interpret them through instincts developed from a lifetime of interacting with other humans.

I began noticing this pattern repeatedly at robotics demonstrations and gatherings. When a humanoid entered a room, the first few seconds were always revealing. People watched closely, not just for technical capability but for behavioral cues. They observed how the machine oriented its body, how it adjusted its steps, and how it positioned its arms while interacting with objects. If those movements aligned with familiar human patterns, the room relaxed quickly. Conversations resumed, and the robot became part of the environment rather than the sole focus of attention.

The same reaction appeared across different audiences. Engineers analyzed performance metrics, but even they instinctively reacted to posture and rhythm before discussing control algorithms. Investors watching demonstrations often commented on how natural a robot appeared when navigating a crowded space. Visitors encountering humanoids for the first time frequently evaluated the machine in simple human terms, asking whether it seemed aware of its surroundings.

These reactions illustrate how deeply human perception influences technological adoption. People interpret physical movement through patterns learned from other people. Body orientation signals direction. Arm movement signals intent. Changes in pace signal caution or urgency. Humanoids communicate through the same visual language simply by sharing similar physical structure.

This does not require perfect imitation. In fact, perfect imitation would likely create discomfort rather than trust. What matters is that the machine communicates intent through motion that people can read easily. A humanoid turning its torso slightly before stepping forward signals direction in a way that observers recognize immediately.

A stable hand position during an object handoff signals readiness in a way that feels natural.

Designers of humanoid systems often refer to this property as legibility. A legible robot allows people to understand what it is doing without needing explicit instructions. That understanding reduces hesitation during interaction. When individuals can predict how a machine will behave, they adjust their own actions with confidence rather than caution.

Legibility becomes particularly important in environments where people and machines share space continuously. Hospitals, airports, and logistics facilities all involve dense human activity. Workers carry objects, change direction unexpectedly, and communicate through gestures rather than formal instructions. A robot operating in these environments must move in ways that fit the rhythm of that activity.

Humanoids provide a strong starting point because their bodies naturally align with the same movement patterns. A machine with two arms and a vertical posture occupies space similarly to a person. When it slows near intersections or angles its shoulders while passing through narrow corridors, those adjustments mirror human navigation behavior.

The result is not simply aesthetic familiarity. It is operational compatibility. Workers navigating the same environment already understand how bodies move through space. A humanoid following those patterns integrates into existing routines with far less disruption than a machine that moves unpredictably.

The physical design of humanoids therefore acts as a compatibility layer between advanced robotics and the built world. Instead of forcing industries to redesign their environments around automation, humanoids adapt themselves to infrastructure that already exists.

This compatibility also extends to tools and objects. Nearly every tool in common use today was designed for the human hand. Handles, levers, and control panels follow conventions that assume a certain range of motion and grip orientation. A robot designed with similar reach and manipulation capability can interact with these objects directly rather than requiring custom interfaces.

That ability dramatically expands where machines can operate. A humanoid capable of grasping standard tools can participate in tasks that would otherwise require specialized robotic equipment. Even when modifications are eventually introduced to optimize workflows, early deployments can begin without waiting for those changes.

Observing these dynamics across multiple environments gradually reinforced an important realization for me. The humanoid form is not simply a design choice driven by aesthetics or science fiction imagery. It is a practical response to the geometry of the world humans have built.

Every staircase, countertop, workstation, and storage rack reflects assumptions about how people move and reach. Those assumptions have accumulated across centuries of architecture and industrial design. By aligning with them, humanoids inherit access to environments that already exist on a global scale.

The implications become even clearer when considering how technology spreads through industries. Adoption rarely happens all at once. Early deployments appear in specific locations where the benefits are easiest to demonstrate. Successful pilots lead to replication in similar environments, gradually expanding the footprint of the technology.

A machine that requires significant environmental changes slows that process because each new deployment demands additional preparation. A machine that fits naturally into existing environments can expand much more quickly.

Humanoids therefore benefit from an unusual form of scalability. Their compatibility with human infrastructure allows them to move between environments with minimal adjustment. A robot trained in one warehouse or hotel may operate in another with only small changes in parameters such as navigation paths or object placement.

This portability strengthens the economic case for the technology. Companies evaluating humanoid systems often focus on performance metrics, but they also consider how easily the machines can be deployed across multiple sites. The less infrastructure modification required, the easier it becomes to scale operations.

Over time, this compatibility begins to influence expectations about how robots should behave in human spaces. As people encounter humanoids more frequently, they develop clearer mental models for how these machines interact with their surroundings. That familiarity reduces hesitation and accelerates acceptance.

Humanoids are likely to follow a similar path. Early encounters will attract attention and curiosity, but repeated exposure will transform them into ordinary elements of the environment.

What enables that transition is not just technical capability but the alignment between machine design and human expectation. Humanoids communicate through motion and posture that people already understand, allowing them to integrate into spaces without demanding new behavioral rules.

In this sense, the form factor of humanoids serves as a bridge between advanced artificial intelligence and the everyday environments where that intelligence will eventually operate. It provides a physical language that humans can interpret instinctively, reducing friction during the earliest stages of adoption.

As the technology continues to evolve, the capabilities of these machines will expand dramatically. Yet the underlying principle will remain consistent: The closer robots align with the structures and behaviors of the human world, the easier it becomes for that world to accept them.

That acceptance is what ultimately determines whether humanoids remain laboratory demonstrations or become part of the infrastructure that shapes daily life.

The Co-Evolution of Humans and Technology

Humanoids enter the world at a moment when people are already surrounded by intelligent systems. Phones predict intent. Cars sense lanes. Software finishes sentences. The idea that machines adapt to us is part of daily life. But intelligence now has a body. Once machines began to move through physical space, the familiar loop between humans and tools reappeared, only faster and more visible.

Every major shift in technology has followed the same rhythm. People make tools that solve constraints. The tools change what people do. People adjust their habits, expectations, and spaces. Then the next version of the tool is shaped by those adjustments.

Fit gives humanoids a head start, but it's only the beginning.

The Loop in Practice

The real movement begins when humans and machines start shaping each other through practice.

Even in controlled demonstrations, the loop is visible. People watch how the machine moves, then unconsciously adjust how they stand or walk. They create spacing that did not exist a moment earlier. They form natural arcs to give the machine a clear lane. None of this is taught. It is instinct. The environment itself shifts because the presence of a humanoid nudges behavior.

As more people gather, the machine responds to them as well. It slows its pace when the crowd thickens. It angles its torso to telegraph direction. It adjusts its shoulder height slightly when someone steps close for a photo. These small movements feed back into the room. People copy them. People trust them.

At a retail store where a prototype was shown, staff gathered at the edges to observe how the robot handled narrow aisles. Customers paused mid-step, read the machine's posture, and repositioned themselves without thinking. In a university lab, students moved from casual curiosity to deliberate analysis within minutes. The humanoid adjusted as well, learning how to navigate through scattered chairs and unpredictable foot traffic. Each side influenced the other.

The speed of humanoid integration has become visible. When drones first entered public spaces, people had to learn how to behave around them because nothing about a drone was familiar. Autonomous cars created a similar moment. People stared at the driver's absence and wondered whether to make eye contact with the sensor. With humanoids, the learning curve is shorter because people already understand bodies that move like theirs.

That loop shows up in physical details first, eventually spreading to communication, design, and operation. A building decides where a machine should wait before entering a busier area. A team defines a comfortable handover height. A store adds a small decal near a doorframe to mark the spot where the robot pauses before crossing. These adjustments are small, but they compound.

The same applies to the robot. A preferred passing side becomes part of its default logic. A common shoulder angle before turning becomes part of a movement library. A clutch point near a narrow section of an aisle becomes part of the navigation policy. A controlled pause at the edge of a crowd becomes part of its social grammar. All decisions shape how the machine evolves.

This loop extends beyond movement into how people communicate with the platform. At early events, people tried to talk to the robots the way they speak to voice assistants. They issued long commands, waited for conversation, or treated the robot as a character. Over time, the tone shifted. People realized that humanoids respond best to clear, concise cues: simple phrases and simple gestures. The machine adapted, too, learning to acknowledge with a head tilt, a small shift in posture, or a quick visual signal.

Early humanoids often shipped with a uniform height and arm span. As builders collected more real-world data, they made adjustments: slightly narrower shoulders to navigate crowded aisles, slightly softer head shapes that reduce intimidating silhouettes, wrist ranges tuned to match common human patterns. These refinements were not about engineering alone. They came from how people reacted when a machine entered their space.

Early screens were filled with icons and text. Later versions became cleaner: action, next action, how to stop. That pattern came from watching how people look at machines in real environments and how much information they can process at a glance without breaking their own workflow.

How the Loop Shapes Scale

Companies evolve alongside this. Buyers write requirements that mirror human expectations instead of technical specifications. Instead of asking how many joints a robot has, they ask how predictable its behavior feels during a crowded demo. Instead of asking about torque, they ask how well it communicates intent in fast environments.

Service models adjust too. Traditional robotics depended on long integration cycles, custom deployments, and support tickets. Humanoids push for something more fluid. Behavior tweaks can ship with the same rhythm as software updates. Minor improvements travel from one site to another through light skill packs. Builders and operators share clips, notes, and feedback loops that feed directly into model updates. The support structure becomes collaborative rather than hierarchical, and the loop between use and improvement tightens.

Co-evolution also shapes how companies think about cost and scale. Early units are expensive and limited. Adoption starts small. As behavior improves, comfort increases. As comfort increases, deployments expand. Better products attract more buyers. More buyers raise production volume. Costs fall. When acceptance grows, economics improve, and the loop is direct and easy to observe.

Societal patterns around humanoids signal a broader shift. At public events, children react with natural curiosity, while adults show analytical caution. At corporate gatherings, executives mix fascination with practical questions. At design studios, teams focus on posture, expressiveness, and emotional tone. At tech expos, the reaction is often a flash of excitement followed by a moment of deep evaluation. These reactions come from different worlds, but together they form a coherent signal: Society is adjusting faster than expected, and a new kind of cultural literacy is forming.

People are learning to read the intent of a moving machine, the same way they read a cyclist entering an intersection or a person approaching a door. They understand the cues. This literacy is still forming, but once it stabilizes, it will accelerate everything.

That stability will not come from one company or one design. It will come from a steady cycle of interaction, feedback, and refinement across sectors. Over time, a shared set of norms will surface, not through committees or standards groups alone, but through practice repeated at scale.

This loop explains why humanoids are not a novelty phase, but a progression. It explains why certain designs work better than others. It explains why the timing finally aligns. And it explains why the movement feels bigger than a single product line or company.

Co-evolution is not a side story. It is the mechanism that determines what survives. The world is not reorganizing itself around robots.

Robots are learning to fit into the world as it is. In return, people are making light adjustments that improve safety, clarity, and performance.

Once I understood how people respond to intelligent machines in real life, it became clear that the timing was no accident. Something larger was aligning, and that alignment is where the broader story begins.

The world was not waiting for a robot. It was waiting for convergence, and the convergence finally arrived.

2

The Missing Ecosystem I Had to Build

I HAD LIVED through the autonomous vehicle industry from its earliest research stages to its first attempts at commercial maturity. It takes a long time for a new field to become real, and it depends on timing, structure, and the systems around the technology. When I began looking closely at humanoid robotics, that experience shaped how I read the landscape. The pattern felt familiar, but the gap was larger.

The more time I spent inside labs and around early builders, the clearer that gap became. There were pockets of advanced work, but almost no shared pathways, no aligned incentives, and no consistent support structures. Founders were solving important pieces, often with limited resources, and doing it alone. The absence of an ecosystem was obvious to me because I had already seen what happens when a field grows without one. Momentum can build, but it cannot carry itself without the foundation that links ideas, capital, talent, and adoption.

That context shaped how I understood that the technical steps forward were real, and they mattered. But they did not yet change the underlying reality.

Before the Humanoid Momentum Took Shape

Before humanoids gained real momentum, Silicon Valley and other innovation hubs had a familiar intensity. Founders were building fast, experimenting with new motion systems, custom actuators, early control stacks, and compact sensor arrays. The culture rewarded taking big swings, and the people entering the field were comfortable working in complex, unstructured environments. The work was bold, and, in many cases, ahead of what the surrounding ecosystem was prepared to support.

The progress was real but scattered. The pieces were strong on their own but were isolated in silos. What they lacked was alignment.

Isolated Breakthroughs

In the early days, one team might unveil a breakthrough in locomotion, showing a robot that held balance through dynamic shifts that would have seemed impossible only a few years earlier. Another team might show a perception stack that could identify human movement with impressive accuracy. Many of the experiments reflected the classic Silicon Valley mindset: Build something ambitious first, then refine it later. Founders were not waiting for perfect environments or tightly defined use cases. They were exploring the boundaries of what humanoids could physically do. Willingness to try mattered almost as much as the results themselves.

Prototypes were rolling out every few months. People were testing new joints, new compute modules, and early AI extensions that hinted at what embodied intelligence could become. Energy systems improved enough to support longer tests. Edge compute modules began to reach performance levels that could handle more complex models. These steps slowly showed the direction of travel, even if the broader environment had not yet caught up.

But this same openness also meant the field lacked the structure needed for coordination. The prototypes were not designed to integrate with each other. A company exploring advanced balance did not always consider the constraints of manipulation. A team working on manipulation did not always think about the realities of mobility.

Everyone was working on critical components, but without a shared path, the advances remained disconnected. This lack of connection limited how far any single team could go, even when its own work was strong. The foundations were there, but the world outside the labs saw something entirely different.

The Gap Between Progress and Readiness

Outside the labs, the field still appeared early. Observers saw impressive demonstrations, but they were not yet tied to clear deployment pathways. There were also misconceptions about the field. Many observers believed humanoids were still too early to consider seriously. The surrounding infrastructure was not yet formed. They saw impressive demos but not the broader scaffolding that would support cost reduction, integration, reliability, and long-term operation. Without that scaffolding, the demos stayed demos. And from the outside, it appeared that the field was advancing more slowly than it truly was.

The mismatch between technical progress and structural readiness shaped how potential customers viewed the field. Companies in logistics, manufacturing, healthcare, and service industries were accustomed to automation that was specific, predictable, and easy to contain. They were familiar with warehouse robots, robotic arms, or fixed systems designed for narrow workflows. Humanoids, by design, were more flexible and more capable, but they required deeper integration and a broader understanding of how they would operate in real human environments. At the time, that intersection was not yet clear. The use cases were intuitive, but the pathways to get there were not.

Venture capitalists saw that the timing did not yet align. In those early years, the signals were mixed. Hardware was improving, but unit costs were high. AI was advancing, but still brittle in real-world variance. Sensors and actuators were becoming more capable but not yet optimized for scale. And market acceptance, while growing, did not yet reflect a strong pull from operators who would ultimately deploy humanoids at scale.

Humanoids demand synchronization across domains that typically evolve independently. Hardware maturity has to match AI capability. Cost curves have to match investor expectations. Use cases have to align with market acceptance. And the appetite for capital has to match the timeline for development.

Scattered Efforts Without a System

As innovations continued to rise, their placement seemed increasingly disconnected. Startups with an interest in mobility were building platforms that could balance, recover from small disturbances, and navigate controlled indoor environments with growing confidence. Their focus was almost entirely on motion. Every cycle of development improved stability, foot placement, torque control, and gait planning. Founders built custom actuators and gear trains or adapted high-power units from adjacent industries. The work was impressive, but locomotion breakthroughs did not automatically translate into progress in manipulation or perception. It solved an important problem but solved it alone.

Other founders concentrated on the upper body. They focused on dexterity, finger control, wrist compliance, and stable grasping strategies. Manipulation was its own frontier, and many startups approached it without a strong link to mobility. Some optimized fine control without considering how the hands would function while the robot moved. They built elegant solutions that performed well on benches and rigs but lacked full-body control integration.

Perception was evolving along a separate track. Founders working on computer vision and lightweight AI models were building early systems that could interpret human motion, detect objects, and estimate depth with reasonable accuracy. Some of these systems were adapted from work in mobile robotics, others from mixed reality tools. Perception teams did not always have access to the motion hardware, so they tested models on fixed platforms or simplified datasets.

Actuation engineers pushed hard on compact designs, high torque density, and better thermal performance. These developments were critical for making humanoids practical, yet they were not always built with specific full-body control strategies in mind. Hardware teams optimized for efficiency, weight, or torque, while control teams worked on inverse kinematics and stability using whichever actuators happened to be available. Integration was possible, but only after the fact, and often at significant cost.

Compute advanced on its own timeline as well. Edge devices gained power and efficiency at a rapid pace, allowing builders to run increasingly complex models onboard. But compute strategy was not coordinated with the hardware design or control architecture. One company might use a high-performance module that requires careful thermal management, while another might build around lower-power components to extend runtime.

AI models during this period were improving, but they were not yet built for the kind of broad generalization humanoids required. Early perception and policy models were narrow and brittle. They worked well when the environment matched the assumptions built into the training data. They struggled when those assumptions changed. Without a common dataset and large-scale learning loops built on real-world experience, AI remained a limiting factor. It added capability but did not unify the field.

Data scarcity sat quietly at the center of this fragmentation. Every team needed data, but no one had enough of the right kind. Teleoperation offered realism and direct human demonstrations, but it was slow, inconsistent, and expensive to collect at scale. Simulation generated large volumes of synthetic experience, but it did not match the complexity and variation of real environments. Neither could fully support the kind of robust generalization humanoids would eventually need. And because these efforts happened independently, the gains were contained within individual projects, not shared across the field.

The independent industries worked in parallel, each solving a piece of the puzzle, but none speaking the same language. The systems worked in isolation but failed in combination.

The Cost of Independence

This fragmentation shaped how prototypes behaved in practice. A robot that excelled at balance in one startup's environment might fail entirely when placed on slightly uneven terrain or when asked to manipulate objects while moving. A manipulation strategy that performed well in a controlled setting could fall apart when integrated with a different motion stack. A perception system that detected human behavior accurately in a laboratory might misinterpret the same signals in a warehouse, hospital, or home.

The absence of shared frameworks reinforced this pattern. Many founders built their own software stacks, control libraries, simulation tools, and training pipelines. These systems reflected the unique strengths and constraints of each group, but they did not connect easily with one another. If a startup wanted to adopt another team's perception model, it often needed to rewrite significant parts of its own control system. If a lab developed a new balancing algorithm, integrating it into a different platform required modifications that few teams had the bandwidth to pursue. Collaboration was possible, but it was not seamless.

That uncertainty slowed everything downstream. Without confident early adopters, companies could not design deployment road maps. Without road maps, investors could not form durable theses. Without investment, founders had to choose narrow slices of the problem to work on. It was a reinforcing cycle: Fragmentation limited adoption, which limited capital, which limited focus, ultimately deepening fragmentation.

Yet each team's progress contributed to the field's long-term trajectory. The experiments in motion, manipulation, perception, and AI created building blocks that would later become essential. They revealed the boundaries of what was possible at the time and showed where the next breakthroughs needed to occur. Talent, ambition, and forward movement were not the limiting factors. The limiting factor was that these contributions were not part of a coordinated system.

As the next phase approached, the scattered efforts began to signal something new. The work was still independent, but the edges were starting to touch. Locomotion platforms improved enough to support more complex manipulation strategies. Perception models became flexible enough to handle broader variations. Actuators reached performance levels that could support full-body coordination. Compute modules advanced to the point where more sophisticated AI could run onboard. These shifts did not create momentum on their own, but they revealed the outline of what a more connected field could become.

The pieces did not yet fit together, but they were moving in directions that hinted at eventual convergence. The challenge was to connect these breakthroughs into a unified path. Yet from the outside, many observers still failed to recognize what was happening.

Why the Signals Were Misread

One reason humanoid development remained underestimated for so long is that the signals of progress were distributed across many disciplines that rarely spoke to each other in a unified way. Robotics engineers, artificial intelligence researchers, hardware manufacturers, and venture investors were each observing parts of the same transformation, but they were interpreting those signals through the lens of their own field. Because the advances were occurring in parallel rather than inside a single industry narrative, the broader trajectory remained difficult to recognize.

Within robotics laboratories, progress was visible but measured. Improvements in locomotion, manipulation, and balance emerged through years of detailed engineering work. A robot capable of maintaining dynamic stability while walking represented the culmination of long research cycles involving control theory, actuation design, and sensor integration. An improved actuator might deliver better torque efficiency or reduced weight, yet from the outside it appeared incremental rather than transformative. Engineers working inside the field understood how meaningful these improvements were, but the cumulative effect of those advances was not always visible to observers who encountered the technology only occasionally.

Artificial intelligence research was progressing along a different path. Breakthroughs were occurring largely in software and data rather than in physical machines. Machine learning models became increasingly capable of recognizing patterns, interpreting images, and learning from large datasets. Reinforcement learning methods allowed systems to improve behavior through repeated training rather than through explicit programming. These advances were rapidly expanding what machines could perceive and interpret, yet the implications for robotics were not always immediately apparent. Many AI researchers focused on digital applications such as language processing, search systems, and recommendation engines, and their work was not always framed in the context of embodied intelligence.

Hardware development represented another layer of progress that unfolded somewhat independently. Semiconductor manufacturers and embedded systems engineers were steadily increasing the computational capacity available within compact devices. Sensors capable of measuring force, motion, depth, and environmental context became smaller and more reliable. Improvements in batteries and power electronics allowed mobile systems to operate for longer periods without interruption. These developments were essential for robotics, yet they were rarely discussed through that lens because their applications extended across many industries.

Each field therefore interpreted progress differently. Robotics engineers saw the complexity of mechanical systems that still required refinement. Artificial intelligence researchers saw increasingly capable models but often had limited exposure to the constraints of real-world machines. Hardware companies focused on performance improvements without necessarily linking those gains to humanoid mobility or manipulation. Because the advances were distributed across these separate domains, the broader pattern remained difficult to see.

From outside these specialized communities, the field appeared even more fragmented. Media coverage of robotics often focused on individual demonstrations rather than on the engineering trajectory behind them. A video showing a robot walking across a stage or performing a simple task could appear impressive, but it did not necessarily communicate the technical progress required to make that moment possible. Demonstrations were frequently interpreted as isolated experiments rather than as signals that the underlying capabilities were improving steadily.

This fragmentation created a perception gap that persisted for many years. The people closest to the technology could see that meaningful advances were occurring across multiple domains, but those advances were distributed in ways that prevented the broader market from recognizing the pattern. Observers evaluating robotics through occasional demonstrations or headlines encountered a series of disconnected signals rather than a coherent narrative of progress.

The venture capital community often reflects these perception dynamics. Investors evaluate emerging technologies by examining timing, scalability, and market readiness. When a field appears fragmented or uncertain, capital tends to flow toward areas where the path to commercialization is easier to model. Industrial automation offered such clarity because its environments were controlled and its economics were well understood.

A robotic arm designed to weld or assemble components operates inside a carefully structured environment. Its task is specific, repeatable, and measurable. Logistics robots navigating warehouse aisles deliver similar predictability because their routes and workflows are defined in advance. These systems provide clear economic returns and operate within boundaries that engineers and operators can easily manage.

Humanoid robotics did not fit comfortably within that framework. The technology was designed to operate across a wide range of environments originally built for people rather than machines. Instead of performing one narrowly defined function, humanoids were intended to adapt to many different tasks. This flexibility represented a profound opportunity, but it also made evaluation more complicated because there was no single deployment model that captured their potential.

As a result, many observers concluded that humanoids were still too early to consider seriously. The machines appeared promising but incomplete, and the surrounding ecosystem had not yet formed. Supply chains, operational standards, regulatory guidance, and integration partners were still evolving. Without these elements, large-scale deployment appeared distant even though the technical foundation was steadily improving.

Researchers and founders working closely with the technology began to see a different picture as the underlying components matured. Locomotion systems became capable of maintaining stability across increasingly complex movements. Manipulation systems improved their ability to interact with everyday objects that varied in shape and texture. Perception systems expanded their ability to interpret complex scenes using combinations of cameras, depth sensors, and machine learning models.

Artificial intelligence systems were also becoming more adaptable in real environments. Machine learning models trained on large datasets could recognize patterns and behaviors that earlier rule-based systems struggled to capture. Reinforcement learning introduced new ways for machines to refine their behavior through repeated interaction with the world. These techniques gradually reduced the gap between controlled demonstrations and practical operation.

The convergence of these developments suggested that the long-standing barriers to humanoid robotics were beginning to weaken. Recognizing this convergence required stepping back from individual breakthroughs and observing how progress across multiple domains was reinforcing itself.

Technological revolutions rarely emerge from a single discovery. The Internet existed for years before broadband networks and web applications unlocked its full potential. Smartphones evolved gradually until touch interfaces, mobile operating systems, and wireless infrastructure matured together. In each case, separate streams of progress eventually aligned to create a platform capable of transforming entire industries.

Humanoid robotics was approaching a similar threshold. Artificial intelligence was expanding the ability of machines to interpret complex environments. Sensors and hardware improvements allowed robots to gather and process richer information about their surroundings. Actuators, materials, and control systems were improving the fluidity and reliability of motion. At the same time, economic and demographic pressures were increasing demand for automation capable of operating outside traditional industrial settings.

Viewed independently, each of these developments appeared incremental. Taken together, they were reshaping the conditions required for humanoid deployment.

Another factor complicated how observers interpreted progress in robotics. Because robots interact directly with the physical world, their capabilities are immediately visible. Early demonstrations inevitably reveal imperfections in movement, timing, or manipulation. Observers often interpret those imperfections as evidence that the technology remains far from practical use.

This visibility can distort perceptions of progress because robotics systems improve through continuous iteration. Small improvements in sensing accuracy, actuator efficiency, or control algorithms can significantly improve performance over time. Movements that initially appear slow or awkward can become fluid through repeated refinement of both hardware and software.

These incremental improvements are often difficult to recognize until a threshold is reached where the overall system behaves noticeably differently. Once that threshold is crossed, the technology can appear to advance suddenly even though the underlying progress accumulated gradually over many years.

For those closely observing the field, the signals of convergence were becoming increasingly difficult to ignore. The capabilities required for humanoid systems were no longer confined to isolated research projects. They were emerging simultaneously across the disciplines required to support embodied intelligence.

What remained missing was not the technology itself, but the coordination required to connect those advances into a functioning ecosystem. Transformative technologies rarely succeed through isolated breakthroughs alone because they depend on alignment across research, capital, manufacturing, policy, and public understanding.

Without that coordination, progress remains scattered and difficult to translate into widespread adoption. Breakthroughs occur but fail to scale because the surrounding structures required for deployment have not yet formed. Evaluating humanoids purely as individual machines obscured the broader transformation that was beginning to take shape.

The blind spot surrounding humanoid robotics therefore reflected more than technical skepticism. It reflected a structural misunderstanding of how complex technological ecosystems emerge and how progress across multiple domains eventually converges into a coherent platform.

When the Missing Ecosystem Became Clear

The early signals did not come from one place or one type of organization. They appeared across several parts of the industry. Major players began stepping forward and reframing humanoids as something far more serious than a research curiosity. Their announcements shifted the energy in the field. They showed that the idea was no longer theoretical. It was being taken on by groups with the confidence and capability to move advanced robotics into real-world settings.

Soon after, the broader robotics community began expressing interest in humanoids at a speed that felt different from previous cycles. New companies appeared with focused missions. Early-stage teams began publishing progress that showed clear forward movement. Major research institutions released detailed analyses that treated humanoids as a natural extension of embodied intelligence. Industry discussions referenced them more often. Long-running robotics events began giving humanoid developers more space, both on stage and in exhibition areas. They were no longer side projects. They were becoming central to the conversation.

A Global Convergence

At the same time, activity outside the United States was becoming hard to ignore. Large exhibitions abroad began featuring humanoids in prominent ways. Traditional robotics vendors started exploring humanoid platforms. Even established manufacturers, which tend to move cautiously with new form factors, were experimenting with early prototypes. National research groups and industry organizations released their own positioning statements on humanoid systems. The conversation was growing in density and seriousness.

What had once been treated as an ambitious but distant category was suddenly a topic of global attention.

It was the pattern behind the noise that became much more meaningful than the noise itself. The activity was coming from different parts of the world, different parts of the industry, and different types of institutions. It was not one cluster trying to spark interest. It was a distributed shift. That level of convergence rarely happens by accident. It usually reflects a deeper technological or market readiness that people are starting to sense at the same time. Even without comparing regions or assigning credit, the message was clear. Humanoids were entering a new phase, and the movement was becoming too pronounced to dismiss.

My background in autonomous vehicles made these signals easier to interpret. I had lived through the full arc of that industry, from early work inside labs to investment waves, pilot programs, public expectations, and attempts at commercial maturity. In that field, the earliest signs of a coming wave were always scattered and easy to overlook. But when you paid close attention, you could identify the moments when interest stopped being isolated. You could feel when a topic started moving from the edge of industry conversation to the center. When I saw the growing attention around humanoids, I recognized the same pattern. It was not hype. It was coordinated acceleration.

The US Landscape: Talent Without Connection

As I shifted my focus to the landscape inside the United States, I saw extraordinary technical talent working across locomotion, actuation, manipulation, perception, and AI. I saw founders pushing hardware limits. I saw others innovating in control, reinforcement learning, and large-scale demonstration data. Yet for all this depth, the work was still unfolding inside separate pockets. Startups operated on their own. Researchers published important breakthroughs, but without a shared industry anchor to tie the threads together. Suppliers developed components that could be used in humanoids, but without a structure that encouraged common interfaces, shared resources, or unified direction.

Technical fragmentation reflected this more clearly than anything else. Locomotion teams were advancing balance and stability. Manipulation teams were building high-precision hands. Perception teams were creating robust sensing pipelines. But each worked independently. They had their own priorities, their own timelines, and their own incentives. No single team was responsible for connecting the rest. No platform existed to bind them together. Even data pipelines highlighted the issue. Some teams relied on teleoperation, which offered realism but slow collection. Others leaned on simulation, which scaled but struggled with edge-case complexity. These approaches were valid, but they were not aligned or shared.

The more I studied these pieces, the clearer the underlying issue became. The problem was not capability. It was coordination. Humanoids require alignment across multiple domains at once. Hardware needs to reach a point where reliability and efficiency support real deployment. AI needs to mature from narrow behaviors to generalizable autonomy.

Operators need clear pathways for integration. Investors need the confidence that the timelines, cost curves, and market pull justify the capital. Suppliers need stable demand signals. Standards bodies need clarity around safety frameworks. If any of these elements are missing, progress slows. If several are missing at once, adoption stays out of reach.

In the United States, builders created impressive prototypes, but there was no shared platform for the field to compound. Investors were curious, but there was no dedicated capital stack for humanoids and embodied intelligence. Suppliers developed strong hardware, but without a unifying structure that encouraged interoperability. Policy discussions were emerging, but without a central industry group organizing them. The gap was not about innovation strength. It was about ecosystem readiness.

This is when the missing ecosystem became clear. It was not an abstract idea or a theoretical gap. It was a concrete set of structural needs that nobody had stepped in to build. The field needed coordination across capital, standards, policy, industry, hardware, and software. It needed a platform where startups, researchers, operators, manufacturers, and investors could work within a shared context. It needed neutral ground to connect global builders. It needed an anchor inside the United States that could serve as a reliable point of gravity for the entire movement.

No individual company was positioned to take on this role. Their focus had to remain on their own products. They had investors to answer to, milestones to hit, manufacturing to plan, teams to grow, and technology to build. Research groups could contribute knowledge, but they could not architect an industry. Suppliers operated according to their own development cycles. Policymakers could provide guidance, but they needed industry input. Builders could push the field forward, but they could not construct the environment around it. Ecosystem building required a different type of mandate.

It required someone willing to take on the connective work directly.

The more I thought about it, the more I realized that the gap was not going to close on its own. The gap needed to be addressed intentionally by someone willing to invest deeply in the category, not only with time but with capital. It needed to reflect the seriousness of the moment unfolding around humanoids, both in the

United States and abroad. And it needed to give builders a larger structure they could rely on.

The Forces That Finally Aligned

The momentum suggested something new was forming, but momentum alone does not create a market. That started to change. The forces that had once evolved separately began moving in the same direction. The shift was not sudden. It was built gradually, through many small steps that accumulated into a real turning point. What matters is alignment. Humanoids require several domains to advance at the same time, each at a pace that supports the others.

For years, those domains moved independently. Hardware improved on its own timeline. AI matured on a separate track. Operators formed their views based on unrelated pressures. Investors evaluated opportunities according to their own expectations. Suppliers responded to whichever segment offered them the most predictable demand. The result was progress without synchronization.

Technical Foundations: AI and Hardware

The first meaningful shift came from AI. For years, robotics AI faced a limit that many founders understood but could not overcome on their own. Models trained on narrow tasks struggled to generalize. Motion strategies that worked in controlled environments were less reliable in unpredictable settings. Learning from small amounts of human demonstration data did not scale well. Reinforcement learning could achieve strong results, but only after extensive tuning and carefully constructed environments. Robotics needed AI that could adapt across tasks, not just perform one task well. It needed AI that could learn from a broad distribution of experiences, not a narrow one.

The field finally reached a point where these barriers began to loosen. Improvements in perception models made it possible to treat visual information with greater stability. Diffusion models and foundation models in vision and control created new paths for generalizable behavior. Large-scale demonstration pipelines became more realistic to operate, even if imperfect. Simulation quality improved to the point where synthetic data could play a more constructive role in training without drifting too far from reality. Mixing these sources allowed founders to build learning loops that were more efficient and more robust. None of this solved autonomy completely, but it moved AI to a stage where humanoids could begin operating with enough reliability to justify early deployment pathways.

As AI advanced, hardware matched its pace. For a long time, the challenge in hardware was balancing reliability, efficiency, and manufacturability. High-performance actuators existed, but they were often too expensive or required designs that did not scale well. Sensing was accurate, but integration was difficult. Transmission systems produced torque but introduced noise and friction that complicated control. Battery systems were improving but fell short of the uptime needed for real work.

Gradually, hardware started to converge on designs that were both practical and capable. Actuators became more compact and achieved better torque density. Integrated sensing improved stability and control. Thermal characteristics became easier to manage. Power systems extended runtime enough to support early pilots. Joint modules became more consistent and easier to manufacture at small batch sizes. None of these improvements were breakthroughs on their own. They were incremental steps that collectively opened the door for more reliable humanoid motion. This allowed hardware and AI to begin supporting each other rather than evolving in isolation.

Market Conditions: Cost, Operators, and Culture

Technical foundations changed the surrounding conditions. Cost curves began moving in a more predictable direction. Suppliers showed new interest. Manufacturers started exploring the production of key components with higher consistency. Investors became more comfortable with underwriting hardware when paired with a credible autonomy pathway. Founders discovered ways to simplify designs without losing performance. These movements suggested that cost reduction would come not from a single innovation, but from a maturing ecosystem.

Operator behavior shifted as well. Companies facing structural labor shortages began looking beyond traditional automation. They wanted solutions that could adapt to mixed environments. They wanted systems that could operate safely among people. They wanted tools that could learn and update over time. Industrial robots offered clear benefits in structured settings, but humanoids offered flexibility in environments not designed for automation. This meant they started asking better questions. They started identifying tasks that required dexterity, mobility, and autonomy in the same system. Their curiosity signaled that the market pull was beginning to form.

The cultural environment changed alongside these developments. The broader AI wave reshaped how people understood the future of intelligent systems. Embodied AI felt like a natural extension of progress in machine learning. People understood that intelligence in the physical world required different capabilities than intelligence in a digital environment. This reduced skepticism and created space for serious conversations about humanoids. The field was still emerging, but it was emerging into a culture that recognized its relevance.

Venture capital evolved as well, changing field evaluations. Hardware progress created clearer cost trajectories. AI maturity created more predictable timelines for autonomy. Operator interest early on signaled market demand. Global activity created confidence that the field was gaining momentum. These elements supported each other. Investing in humanoids no longer meant betting on one technical breakthrough. It meant recognizing that several areas were moving in a coordinated way. That coordination mattered. It suggested durability. It suggested that the field's progress would continue, even if individual breakthroughs happened at different times.

The Power of Alignment

The alignment signaled a direction for the industry. None of the domains were fully mature. Hardware still had gaps. AI still had limitations. Operators still had questions. Investors still exercised caution. But they were all moving toward the same end state. Their timelines were beginning to overlap. Their expectations were converging. Their constraints were becoming more compatible. This convergence created an environment where progress in one domain amplified progress in another. Improvements in AI made certain hardware designs more feasible. Advances in hardware supported better autonomy. Signals from operators encouraged investors to support the development cycle. Standard work formed because the field was now cohesive enough to require it.

This alignment meant founders became more intentional about designing systems that could integrate with the broader field. They focused more on reliability and manufacturability. They considered how their work would fit into real-world environments rather than only in lab settings. They explored interfaces that other groups could adopt. The field began speaking a more unified language.

The field had reached a point where coordination could accelerate progress instead of slowing it down. The work needed visibility. The alignment made it urgent.

Once the forces aligned, the next phase became clear. The field was no longer defined by separate pockets of innovation. It was defined by a momentum that connected technology, industry, and capital.

When the Path Became Clear Enough to Act

The closer I looked at the field, the more I understood why many people believed humanoids were too complex to succeed in the near term. Struggle appeared in every part of the development cycle. Building a machine that could move with stability, manipulate objects with precision, perceive its surroundings, reason across tasks, operate safely around people, and maintain the reliability needed for deployment required tight coordination across many domains. It was natural for observers to focus on these challenges. Progress had been uneven for years, and advances in one area did not always translate into progress in another. This created a widespread belief that humanoids were promising but not practical.

Humanoid development is defined by tradeoffs. Gains in mobility can reduce efficiency. Improvements in dexterity can affect durability. More advanced perception models require more compute, which affects thermal behavior and runtime. Real-world variability is difficult to capture in controlled tests. Safety expectations add further constraints. Balancing these factors is challenging, and many people concluded that timelines would be long and investment uncertain.

Historical experience backed up the theory. Many robotics categories showed early promise but struggled to reach commercial scale. Prototypes attracted attention, yet adoption remained limited. Those cycles taught people to be careful about believing that a new form factor, especially one as ambitious as a humanoid, could break through where others had stalled.

Autonomy added another layer of concern. Even with advances in machine learning, reliable autonomy in the physical world is difficult to achieve. Models must adapt to inconsistent lighting, clutter, and human behavior. Robots must navigate tight spaces, avoid obstacles, and manipulate objects with varying friction and weight. They must do this with repeatability and safety. Achieving this requires large amounts of meaningful data, which is difficult to obtain. Many viewed this as a long-standing limitation.

These concerns were understandable, but difficulty alone does not define a field's trajectory. What matters is how the core forces evolve. Once those forces start aligning, the difficulty becomes part of a maturing process rather than a barrier.

From Difficulty to Inevitability

The cumulative effect of these shifts changed the way I understood the humanoid trajectory. Inevitability began to take shape. It did not come from optimism. It came from the combined movement of many domains. Humanoids fit environments built for people. They inherit tools, layouts, workflows, and safety practices already in place. They do not require a large-scale redesign of settings. Their morphology supports a wide range of tasks without creating specialized form factors for every function. They unify mobility, dexterity, perception, and reasoning in one platform. Once the underlying technology reached a usable threshold, the value proposition became clear.

Inevitability meant the categories that required flexible, human-compatible interaction would eventually need a platform with these characteristics. Industries facing long-term labor constraints would eventually look for tools that support tasks requiring movement, handling, and contextual understanding. The alignment of AI, hardware, cost direction, and cultural readiness made that trajectory visible.

As these forces reinforced each other, the field's long arc became easier to follow.

Difficulty explained why progress had taken time. Alignment explained why progress would now accelerate. The two ideas coexisted. One required engineering discipline and patience. The other required recognition of the broader forces shaping embodied intelligence. Together, they defined the moment the field reached.

Many people still focused on the difficulty. They saw the challenge of autonomy, the expense of hardware, the demands of integration, and the caution of operators. My vantage point was different. I saw the same challenges, but I also saw the direction of the forces. The robotics community was reaching similar conclusions at the same time. Prototypes demonstrated real capability. Cultural and market conditions were shifting.

The Decision to Act

This recognition clarified the next phase. If the trajectory was moving toward inevitability, the field needed structure. It needed an ecosystem to guide responsible deployment, support builders, connect operators, and help shape the category. That structure would determine how effectively the field could move forward.

The decision then became straightforward. The movement was forming. The talent was here, and the global energy was rising. The pieces existed, but the architecture did not. I had spent enough years building companies, working across robotics and AI fields, and observing how industries form. The pattern was familiar. The opportunity was real. The responsibility felt clear. I knew that if I did not step in, the gap would remain open. So I began building the ecosystem myself, starting with the foundations that would give the field a place to align.

I had already been exploring the idea of starting a venture fund. It was something I had thought about carefully. The plan was not to create another general AI fund. That space was crowded. Every firm in Silicon Valley was competing for similar software deals. The advantage would have been small, and the differentiation limited. I wanted to build something with a clearer foundation and a stronger link to my experience. My work always lived at the intersection of hardware, software, perception, behavior understanding, and real-world systems. I wanted a fund that made use of that background. I just had not identified the right category yet.

The moment aligned with my background in a way that was hard to ignore. My experience in autonomous systems gave me a clear view of how these fields progress. I understood how sensing, perception models, control loops, and hardware maturity interact. I understood the constraints that come from operating in human environments. I understood the long timelines and the need for patient capital. I also understood the importance of standards, safety, operator education, and industry-wide coordination. These were the areas the humanoid field was beginning to need.

It became clear that the venture fund I had been planning could serve a higher purpose. It could be more than a capital pool. It could be a mechanism for shaping the early structure of the field. It could support the companies building the core components of humanoids. It could help bridge the gap between operators and builders. It could accelerate the development of common frameworks and shared knowledge. It could lay the foundation for a broader ecosystem. The fund could become one of the early anchors that the field needed.

The idea came from a recognition that the timing was right and that the field needed coordinated support. The opportunity and the responsibility had become inseparable. The ecosystem would form with or without me. The signals made that clear. But the rate at which it would form, and the direction it would take, were still open. My vantage point made the pattern visible. That recognition made the next step clear.

The Commitment to Build

Choosing to launch a fund focused on humanoids and embodied intelligence was the first formal expression of that decision. It bridged the analysis I had done on the ecosystem gap and the conviction I had developed about the field's trajectory. It was a commitment that required time, effort, and personal capital. It marked the beginning of a greater effort. It was the foundation on which everything else could be built.

As the decision matured, I understood that building an ecosystem would require more than capital. It would require a way to convene the field. It also required a collaborative structure. It required a place where builders, investors, operators, researchers, and policymakers could meet on equal footing. It would require platforms that made the category legible. It would require mechanisms that translated ideas into practice. It would require visibility for emerging work. It would require credibility to match the technical ambition of the field.

This understanding shaped the broader architecture of my approach. The fund would support the builders. The ecosystem work would support the field. The two were connected. They reinforced each other.

This decision closed the period of observation. The next stage required building. It required operational work, coordination, and design. It required creating the platforms, events, frameworks, and alliances to support the field. It required investing in the right companies at the right time. It required connecting builders with operators. It required constructing the early architecture of an emerging industry. This was the shift from seeing the pattern to acting on it.

I found myself moving in the same direction as the humanoid field. The momentum was emerging. The conditions were ready.

3

Building the Infrastructure for a Movement

I REACHED A moment when observing the field was no longer enough. The momentum that had been building in the background was starting to take shape, and the gap that defined the industry became harder to ignore. Progress across AI, hardware, culture, and operator readiness was beginning to align in a way that felt consistent. Founders were solving meaningful problems. Early prototypes were becoming more stable. Operators were starting to ask practical questions rather than theoretical ones. Analysts and industry groups were taking humanoids more seriously. In several regions, interest was rising in visible and coordinated ways. The signals were steady. The direction was clear. Yet the structure around the field remained thin.

This imbalance created a sense of urgency. I had seen a version of it before in the autonomous vehicle industry. A field accelerates, talent gathers, breakthroughs appear, and attention grows, but if the surrounding infrastructure does not form at the right time, the momentum slows. Progress continues, but it becomes uneven. Some work never reaches real-world settings. Other work takes longer than necessary. The gap between technical potential and operational reality expands. Watching humanoids enter this stage with no unified scaffolding made the decision clear. If the field was going to move forward with discipline and clarity, someone needed to begin building the system that supported it.

This realization arrived at the same time I began exploring what a new venture fund could look like. I wanted to build something that matched my experience in complex systems, where software, hardware, perception, and decision-making operate together. I did not want another fund pursuing the same software categories that every investor in Silicon Valley was already focused on. I wanted a direction where my vantage point created real leverage. As I continued studying the signals around humanoids, everything that mattered in a fund strategy pointed in the same direction. The category had the traits that aligned with my background, and the opportunity was taking shape. What it lacked was an ecosystem to support it.

Across the United States, many talented founders and builders were working on humanoids, but their efforts were fragmented. There was no shared place for alignment, no clear structure for engagement, and no coordinated support system. The activity was present, but the connective tissue was not. That gap became difficult to ignore. For a field this complex, momentum alone was not enough. It needed the early scaffolding that could help founders, operators, investors, and institutions move in a coordinated way.

Launching the Venture Capital Fund as the First Piece of Infrastructure

The combination made the choice direct. If humanoids were approaching a defining period, then capital alone would not carry the field. It needed a structure that matched the complexity of the work. It needed a way to connect builders, operators, suppliers, and standards groups. It needed a system that could support responsible progress and reduce friction for early deployment. The work had to start with a clear commitment, and the foundation for that commitment began with the venture fund. It would become the first piece of infrastructure in a much larger system that took shape in the months that followed.

The fund was designed for where the real work in humanoids actually begins. The investment strategy followed directly from that reality. The focus was on seed and pre-seed, where foundational choices still determine outcomes and where technical direction matters more than short-term traction. A meaningful portion of the capital was reserved for follow-on investments, because progress in humanoids is nonlinear and founders solving hard problems often need continuity rather than constant reset. The thesis was intentionally full stack. In humanoids, failure rarely comes from a single layer. It emerges at the interfaces between perception, control, hardware design, compute strategy, and real-world environment. Investing narrowly would miss the system-level risks and opportunities that define whether these machines can ever work reliably. This is where many investors struggle. It was where my background provided clarity.

Portfolio construction followed the same logic as the thesis. This was not a strategy built on broad diversification across unrelated bets. In a category as complex as humanoids, understanding compounds faster than coverage. Some investments were designed to mature earlier, particularly in foundational technologies where value could be realized within a few product cycles. Others were made with longer horizons in mind, where system integration, scale, and adoption would take time. Follow-on capital was reserved for founders who demonstrated learning velocity, system-level progress, and increasing alignment with real-world deployment. The goal was not to optimize for early marks, but to support the systems that could compound into enduring platforms over time.

The Investment Thesis

That thesis centered on humanoids, embodied AI, and spatial intelligence as a single, interconnected frontier. Embodied AI refers to intelligence that exists within a physical system and is shaped by interaction with the real world. Unlike disembodied models that operate only on text, images, or abstract data, embodied AI learns through movement, sensing, and consequence. Its understanding is grounded in physics, timing, and uncertainty. In humanoids, embodied AI governs how perception, learning, and action co-evolve inside a body that must balance, manipulate, and recover when conditions change. It is what allows intelligence to be tested not just in simulation or benchmarks, but in lived environments.

Within that frame, spatial intelligence refers to the ability of a system to understand, reason about, and act within three-dimensional environments over time. It sits between sensing and action, enabling a machine to form internal models of space, anticipate how the world changes, and make decisions that remain valid as conditions shift. In humanoids, spatial intelligence determines whether a robot can move through human environments safely, manipulate objects reliably, and adapt when plans break.

Embodied AI provides the body and the learning capability. Spatial intelligence allows it to operate with context, continuity, and intent.

These were not separate categories. They were different expressions of the same problem: how intelligent systems perceive, reason, and act inside real environments. That required investing across the full stack. Autonomy and decision-making. Perception and sensor fusion. Simulation and data generation. Actuation, manipulation, and balance. Compute strategy, power systems, and hardware-software integration. Even enabling layers such as tooling, testing infrastructure, and manufacturing pathways mattered, because they shaped what could realistically be built and deployed.

This was not a software-only approach, and it was not a traditional robotics fund either. It was a systems thesis. The belief was that humanoids would only succeed if intelligence, hardware, and environment were designed together. Software could not be abstracted away from physics. Hardware could not be separated from learning. And deployment could not be treated as an afterthought. The fund was built to support teams working at those intersections, where progress is slower to explain but far more durable once achieved.

Geography was part of the thesis from the beginning. The foundational work in humanoids was never going to emerge from a single region. Some of the deepest research and control expertise lived in Europe. Many of the most capable suppliers and manufacturing partners were in Asia. Much of the autonomy, integration, and capital formation activity was concentrated in the United States. A fund confined to one geography would see only part of the picture. From the earliest stage, it was possible to engage across regions precisely because the category was still forming. Founders were open. Networks were not yet siloed. Collaboration mattered more than proximity.

This approach shaped how capital was deployed. The fund was not built to chase momentum after it became obvious. It was built to move alongside the field as it took shape. Some investments were made during fund formation because the timing demanded it. The work was moving quickly. Strong founders were forming. Infrastructure gaps were already visible. Waiting for a perfectly packaged vehicle would have meant missing the moment when early alignment mattered most. In a category like humanoids, timing is not about speed alone. It is about being present when foundational paths are chosen.

From the start, the fund was intended to do more than allocate capital. It was meant to act as a lens into the field. A way to see patterns early. A way to understand where progress was real and where it was overstated. By working closely with founders at the earliest stages, the fund created a vantage point that revealed not only which companies were emerging, but what the ecosystem itself was missing. That perspective would soon shape everything that followed.

Why This Path

As the idea of building a fund became more concrete, its purpose came into clearer focus. I had been thinking about starting a fund for years, but never in the traditional sense. I was not interested in joining the competition for early-stage AI companies where valuations carried a premium and where every investor in Silicon Valley seemed to be circling the same founders. I wanted a fund that matched the way I work, the way I think, and the way I create leverage. My background was shaped by complex systems, where software, hardware, sensing, and decision-making must operate together inside real environments. I had spent years at the intersection of AI, perception, human behavior understanding, sensor fusion, and real-world robotics. I knew what it meant to solve problems that existed beyond the screen, inside physical settings where uncertainty could not be abstracted away.

There was also something deeper and more personal. I grew up fascinated with robots in a way that stayed with me as I built my career. Long before the professional chapters, I had an early connection to robotics and artificial intelligence. I was drawn to systems that could sense, interpret, and act. Even when the field was quiet and progress was slow, I kept paying attention. I watched traditional robotics move through its cycles of mechatronics, industrial automation, and incremental gains in manipulation and control. I followed the research labs that were experimenting with mobility, dexterity, sensing, and task-level autonomy. I stayed close to the edge of the work, even when there was little commercial momentum behind it. For more than a decade, I watched robotics evolve in parallel to advances in AI, waiting for the moment when the two would finally align.

That moment arrived when deep learning matured, and later when transformers and large-scale AI models made generalization more practical. Suddenly the long arc I had been tracking began to bend. Robotics had always lacked the kind of adaptable intelligence that could lift the field beyond static environments and rigid workflows. Once AI became capable of reasoning across tasks, interpreting unstructured scenes, and learning from broader forms of data, everything I had been following began to converge. The potential of humanoids shifted from long-term speculation to near-term viability, not because the machines were suddenly finished, but because the intelligence needed to control them had crossed a threshold.

At the same time, I was studying the signals that shaped my conviction. From major players making bold commitments, to influential reports, to global activity accelerating across Asia and Europe, it became clear that the field was entering a phase I had seen before. This was the same pattern that preceded the autonomous vehicle wave: rapid technical progress, rising attention, but no ecosystem to bind it together. I knew what happened when a category reached this stage without structure. The momentum would scatter. The breakthroughs would stay isolated. Operators would struggle to understand what was real. Investors would misread the timing. Standards would lag. And deployment would take far longer than necessary.

Past Robotics Struggles

To understand why this moment mattered, it was necessary to look honestly at why similar efforts had struggled before. For decades, robotics sat in an uncomfortable place within venture capital. Hardware was viewed as slow, capital-intensive, and unforgiving. Development cycles stretched across years. Margins were thin. Failures were costly. Software leverage was limited, and generalization was rare. Most robotic systems were designed for tightly controlled environments, where tasks could be scripted and variability minimized. When conditions changed, performance degraded quickly. Venture investors learned to be cautious, and many funds that focused on robotics did not deliver outcomes that justified the risk.

The problem was not a lack of ambition or intelligence. It was structural. Robotics matured in an era when intelligence was brittle and data was scarce. Perception systems struggled outside narrow conditions. Control systems required careful tuning and frequent recalibration. Learning systems could not adapt reliably across tasks or environments. As a result, most robots depended on constrained workflows to function at all. This limited where they could be deployed and how much value they could create. Without generalizable intelligence, robotics remained bounded by its weakest assumptions.

That history shaped how the category was perceived. Robotics was often treated as an engineering discipline rather than a scalable technology platform. Investment strategies favored late-stage industrial automation, niche applications, or services layered around hardware. Funds avoided early-stage robotics altogether, or treated it as an exception rather than a core focus. Capital flowed into software categories where iteration was faster and outcomes were easier to predict. The result was a long period where meaningful progress continued in labs and research groups, but struggled to translate into durable companies.

What changed was not one breakthrough, but a convergence. Advances in deep learning improved perception. Large-scale models made generalization more practical. Simulation and synthetic data expanded the range of scenarios systems could learn from. Compute became more accessible. Tooling improved. At the same time, the environments where robots were expected to operate became better understood. Operators began defining requirements more clearly. Manufacturing processes matured. Supply chains became more capable of supporting complex systems. The gap between what could be built and what could be deployed began to narrow.

Humanoids sat at the center of this shift. They benefited directly from advances in adaptable intelligence, but they also exposed the limitations of partial solutions. A humanoid could not rely on rigid assumptions. It had to balance, perceive, reason, and manipulate in spaces designed for people. This forced a level of integration that earlier robotics efforts could often avoid. It also created a clearer signal of progress. When a system improved, the improvement was visible. When it failed, the failure revealed exactly where work remained.

This convergence altered the risk profile of the category. Hardware was still difficult, but it was no longer blind. Intelligence was no longer static. Learning could persist across tasks. Systems could improve with experience. The long timelines associated with robotics began to align more closely with the expectations of patient capital. For the first time, it became reasonable to build a venture strategy around foundational robotics, not despite its complexity, but because the complexity was finally tractable.

Seen through this lens, humanoids were not a repeat of past cycles. They represented a new phase. One where the historical reasons for caution had begun to erode and where the absence of infrastructure posed a greater risk than the technology itself. That shift reframed the role of capital. The challenge was no longer whether intelligent machines could exist in the real world. It was whether the ecosystem around them would form quickly enough to support responsible progress.

What made the situation different this time was that the opportunity matched every part of my background. Humanoids, embodied AI, and spatial intelligence were the only categories that satisfied all of the conditions I cared about. They required deep technical understanding. They lived at the intersection of perception, control, hardware design, compute strategy, and human interaction. They needed story and vision, but also discipline and engineering. They offered a frontier where my vantage point created real advantage. And most importantly, they required builders who understood that complexity was not a barrier when the underlying forces were moving in alignment.

These connections made the decision clear. The fund needed to be the first step. It would not be a vehicle that chased the companies everyone else was pursuing. It would be a platform that attracted the founders building the future of humanoids. I have always preferred to attract rather than chase. I wanted the fund to become a magnet for the people working on the most difficult problems, not another competitor in the bidding wars surrounding mainstream AI. The fund needed to signal something very specific. It needed to say that the work mattered at the level of long-term impact, not short-term hype. It needed to establish a place where founders felt understood, both technically and strategically.

Slowly, a purpose took shape. The fund would become a lens into the field, revealing where the real work was happening and where progress was durable. It would bring technical rigor and strategic context together while serving as the first anchor in the larger ecosystem the field needed.

Once I began meeting with founders, the clarity deepened. The conversations were revealing. Founders were facing the same patterns I had seen forming. They were navigating uncertainty that came from the lack of structure. They had strong technical instincts, but limited visibility into how operators thought, what buyers prioritized, and which constraints mattered most in early deployments. They were solving hard problems in isolation. They were building promising systems without the benefit of shared context. They were moving forward, but without the connective layer that could help their work scale beyond the prototype stage.

As the thesis sharpened, it became clear that a narrow geographic focus would be a constraint rather than an advantage. The foundational work required to make humanoids viable was already distributed across regions. Control theory, actuation, and safety research had deep roots in Europe. Manufacturing expertise, component reliability, and supply chain depth were concentrated across Asia. Autonomy, software integration, and early commercial experimentation were most active in the United States. Each region held a different piece of the system. None held the whole.

At the earliest stages, this distribution was not a barrier. It was an opportunity. Before categories harden and networks become siloed, founders are more open to collaboration and shared learning. The field had not yet fragmented into regional narratives or competitive camps. Engineers were still comparing approaches. Operators were still defining requirements. Standards had not yet been written. In that window, it was possible to engage across borders with credibility and intent, not as an outsider, but as a participant in a shared problem.

This reality shaped how the fund approached timing. Waiting to invest until every structural element was in place would have missed the moment when foundational paths were being chosen. The category was moving too quickly for that. Strong teams were forming. Core assumptions were being set. Early architecture decisions were locking in constraints that would persist for years. Being present at that stage mattered more than formal completeness. Investing during fund formation was not an exception. It was a response to the pace of the field.

Those early investments were not made in isolation. They were made in context. Each company revealed something about the broader system. Gaps in tooling. Friction in manufacturing. Misalignment between technical capability and operator expectation. Constraints imposed by safety requirements that had not yet been formalized. By engaging early across regions, the fund gained a clearer picture of where progress was durable and where it was fragile. That visibility would have been impossible from a distance.

Cross-region investing also created a different kind of discipline. It forced comparisons. Approaches that worked in one environment were tested against assumptions from another. Hardware decisions were evaluated alongside software strategies. Research-driven systems were measured against operational realities. This constant cross-checking reduced blind spots and exposed weaknesses earlier, when they were still addressable. It also surfaced unexpected strengths. Builders working on narrow components turned out to be foundational. Others pursuing broad visions revealed hidden dependencies that needed support.

Over time, this approach helped connect lessons across regions before the field fully matured. Technical progress could be viewed alongside deployment realities in different operating contexts, creating a more complete picture of where the category was heading. By the time humanoids began drawing broader attention, much of the groundwork had already been laid.

Founders were not just looking for capital. They were looking for clarity. They wanted to understand how their decisions would play out when their robots entered actual environments. They wanted to know how operators evaluated risk, where they showed caution, and where they showed willingness to adopt. They wanted to understand how markets would form around humanoids, even in early stages. These discussions created a vantage point that did not exist anywhere else. They allowed me to see patterns across the field. They revealed where founders converged, where they diverged, and where the industry needed shared understanding.

Through this lens, it became clear that companies alone could not build the momentum the field required. Even the best founders were carrying more weight than any one group could manage. They were building hardware, developing autonomy, engaging operators, educating investors, and navigating the uncertainty of a shifting category. They needed a structure that amplified their work and gave it a place to connect with the rest of the field. Without that structure, every company would be forced to educate the market independently, explain safety logic repeatedly, engage standards groups alone, and build supplier networks from scratch. This was slow, costly, and inefficient. The field needed a shared foundation.

This recognition shaped the decision. If the ecosystem did not exist, then part of the work had to focus on creating the connective layer that would support the category. The fund was not the end. It was the beginning. It would identify the founders shaping the future of humanoids. It would support them. It would connect them. And it would serve as the first node in a broader system that the field needed.

It was never just about capital. It was about shaping the conditions in which humanoids could become real and building the first anchor in a field that needed coherence. It was about setting the stage for the ecosystem that would come next.

Everything that followed grew from this foundation. The fund was the first piece of infrastructure. It created the vantage point that revealed what the ecosystem needed next. It gave the work its starting point. And it set the path for a larger system that would soon take shape around it. That vantage point made the next gap obvious, not in capital, but in coordination.

The Master Plan Behind the Ecosystem I Built

As the fund took shape, the broader architecture of the ecosystem began to come into view. None of the gaps I observed existed in isolation.

They were connected in ways that were easy to miss unless you had spent time inside complex systems where software, hardware, sensing, and human behavior converge. The field was gaining speed, and talent was flowing in, but the structural layers that help an industry move forward together were missing. If the goal was to build real momentum, the work could not stay limited to capital. It had to reach across several dimensions at once. Each layer depended on the others, and without that coordination, progress would remain scattered.

What emerged was a master plan, not in the sense of a rigid blueprint, but as a coherent structure that reflected how the field would need to evolve. It helped clarify the work ahead and revealed where the early foundations needed to be built. The plan came together by mapping each gap to a corresponding pillar of support. Each pillar addressed a different constraint, but none operated independently. Progress in one strengthened the others, and weakness in one slowed the system as a whole.

The structure that emerged was built around several interconnected layers. Capital would support founders solving foundational problems. The Summit would create alignment and shared context. Standards and governance would reduce uncertainty around deployment. Regional organizations would create continuity between major gatherings. Cultural initiatives would help shape public understanding. Policy would provide long-term coordination. Each layer addressed a different challenge, but together they formed a system designed to help the field mature with greater coherence.

The Builders Who Revealed the Gaps: The First Layer of Insight

The fund became the entry point to the field. It surfaced founders working on the hardest problems in humanoids, embodied AI, and spatial intelligence. It allowed me to engage with them early, understand their technical logic, and help them refine their approach to the real conditions that would shape adoption. These conversations revealed insights that were not visible from the outside. They showed where founders converged in their assumptions, where they diverged in their technical paths, and where the industry lacked shared clarity.

But what mattered most was that these insights were not theoretical. They came from the hands-on work of people who were building machines that had to navigate real environments. They reflected the constraints and tradeoffs that shaped early deployment. They revealed the technical and commercial realities that the field would need to address collectively.

I understood quickly that these insights were too valuable to stay private. They needed a place to circulate. Founders needed to hear from one another. Operators needed to understand what was technically feasible. Investors needed a shared foundation for evaluating risk. Without a way to share these insights, each company would be forced to learn the same lessons alone, often at significant cost. Momentum would slow. Fragmentation would deepen.

Company building was not only about investing. It was about creating the first signal that the category mattered. It was about setting a tone that the work deserved discipline and seriousness. It was about building trust with founders and showing that someone understood the complexity of what they were trying to do.

The more these conversations accumulated, the clearer it became that the insights they generated needed a place to circulate.

The Humanoids Summit: The Center of Gravity for a Global Movement

The second layer emerged naturally from the first. As conversations with founders expanded, it became clear that the field lacked a central place where builders, operators, researchers, and investors could learn from one another. Robotics and AI had many events, but none reflected the structure humanoids required. Most were too broad, spanning industrial automation to academic research. Others were too commercial or too general to support the level of alignment this field demanded.

Humanoids required a different kind of environment. One where approaches could be compared directly. Where assumptions could be challenged early. Where progress could be examined with transparency and rigor rather than filtered through marketing or isolated demonstrations.

This was not about visibility. It was about coordination.

Without a shared place to exchange insight, the field risked repeating mistakes, misreading progress, and moving at uneven speeds across regions. The work was becoming too interdependent to remain distributed without a central point of alignment.

The implication was clear. If the field was going to move forward with coherence, it needed a place where its most important voices could meet.

The Summit did not begin as an event idea. It took shape as the answer to a structural problem the field could not solve on its own. Humanoids were gaining momentum. Builders were advancing autonomy, manipulation, and control. Researchers were producing insights that would shape the next decade. Operators were preparing for early deployment. Investors were watching the signals rise. Yet all of this activity had no shared center. The field was moving, but the movement had no place to collect its energy.

The more time I spent inside these conversations, the clearer the need became. The work was too distributed and too interdependent for progress to remain fragmented. Humanoids required a platform that reflected the seriousness of the work and the scale of what was coming. They needed a place where alignment could form, where clarity could emerge, and where the people shaping the category could understand one another's direction.

This is why the Summit was built. It was not designed as a conference. It was designed as infrastructure. It became the connective layer that gave humanoids a voice, a rhythm, and a global center of gravity. Understanding this gap is what led me to begin outlining the structure the field needed and the platform that could hold it.

Establishing the Humanoids Summit

The idea became real as I met more of the people responsible for the early foundation of humanoids. Each conversation gave me a closer view into the work beneath the surface. The builders refining perception, control, and dexterity. The researchers exploring new approaches to learning and generalization. The operators testing assumptions about reliability, safety, and integration. The investors trying to understand what was real. What stood out was not only the progress itself, but how little of it was visible across the rest of the field.

My experience in the autonomous vehicle era made this pattern impossible to ignore. That field produced breakthroughs that shaped an entire decade, but the absence of a unifying structure slowed its long-term acceleration. Builders learned lessons privately instead of collectively. Operators struggled to know what was real. Investors interpreted signals without consistent reference points. Policymakers lacked clarity on how to engage. Progress was real, but the absence of a shared center created friction and delay. A movement cannot form without alignment. A category cannot mature without context.

As I spent more time inside humanoids, I saw the same early risk. Builders already had more capability than the world realized. Researchers were generating insights that could redirect entire approaches to autonomy. Operators were asking informed questions that signaled readiness for real trials. Global interest was rising. Several regions were accelerating. Yet the structure that allowed all of this to inform one another did not exist. Without it, humanoids could become a set of strong but disconnected efforts.

This realization did not arrive dramatically. It built slowly. Every part of a humanoid system depends on another part. Perception shapes control. Control shapes manipulation. Manipulation shapes reliability. Reliability shapes trust. Trust shapes deployment. Deployment shapes investment. No single group could move the category forward alone. The interfaces between these efforts needed to strengthen early, before complexity took over.

Humanoids needed a central node. A place where the most important builders, researchers, and operators could be in conversation.

Creating that node required treating ecosystem infrastructure as seriously as any company investment.

Before I deployed capital into any company, I treated the Summit as the first investment. A fund that aimed to shape a new category needed more than deal flow. It needed a structure the ecosystem could build on. Creating a platform where the field could see itself was not separate from the fund's strategy. It was the groundwork that made every later evaluation sharper. Without a shared center for builders, operators, and researchers to meet, any investment would have been made with incomplete information. The Summit became the starting point, the infrastructure that allowed the fund to invest with discipline and clarity.

A format where the category could reflect its seriousness without becoming theatrical. A stage where the work could be visible in its full depth rather than diluted across broad robotics or AI events. A platform that treated humanoids as a field of its own, not a side topic inside someone else's agenda.

Once this became clear, I evaluated the landscape of existing conferences. Many were excellent, but none matched the requirements of a category as complex as humanoids. Robotics conferences were too broad. AI conferences were too abstract. Automation events focused on industrial workflows that did not map cleanly onto humanoid systems. Some included humanoids as a topic, but none could serve as the center of gravity for a field that needed coherence at a critical moment.

The Summit carried a simple and consistent theme from its earliest blueprint. It focused on three pillars that reflected the real conditions required for humanoids to succeed. Mass production as the force that would turn prototypes into accessible machines. Commercialization as the bridge between promising technology and real operators with real needs. Capital allocation as the mechanism that determined which paths could scale. These themes stayed constant across every edition. They served as a reminder that the category would be shaped not only by technical breakthroughs, but by the practical systems that allowed those breakthroughs to reach the world.

The gap was obvious. The field needed a dedicated platform. Something focused and structured. Something built for the people doing the real work. Something that matched the complexity of humanoids but helped the field think together. Something grounded in rigor, not hype. A place where founders, operators, researchers, investors, and policymakers could meet with intention.

With that intention, the outline for the first Summit took shape. The event needed to send a signal that humanoids had entered a new phase, not in a speculative way, but in a disciplined one.

That foundation shaped everything that followed. The Summit had to become the environment where the category recognized itself, a place where the field could gather with clarity, seriousness, and optimism grounded in real work.

The First Edition

The first edition took place in Silicon Valley. The location mattered. Silicon Valley had always been a place where new technologies gained their earliest form. But humanoids required something different. This was not about software momentum or abstract research. It was about building intelligent physical systems that operate inside real environments. The Summit needed to reflect ambition, but it also needed to reflect discipline. It needed to be shaped by people with a deep understanding of autonomy, perception, actuation, and deployment. It needed to carry the tone of a field ready to move forward.

The inaugural Summit gathered the people who were already shaping humanoids long before they were widely visible. Many had spent years working quietly. Some were building early prototypes. Others were refining autonomy stacks. Some were focused on actuation, manipulation, sensing, or dexterity. Operators who had been watching the category from a distance finally had a place to see the emerging landscape. Investors who had been tracking the signals without full clarity could now understand the structure forming beneath the work.

From the moment the doors opened, it was clear that something new had entered the world.

The audience reflected that shift. Attendees came from more than forty countries, representing every major region influencing the future of robotics and AI. Builders, researchers, operators, and policymakers traveled across borders because humanoids were no longer a regional effort. They were becoming a global discipline that required global alignment.

Companies across the entire humanoid stack were present, from component suppliers developing motors, gearboxes, batteries, and sensor arrays, to simulation teams working on training and validation, to hardware groups refining actuation and system design. Perception and autonomy teams showed advances in vision, mapping, control, and learning. End to end system integrators stood side by side with full humanoid platforms that were being shown publicly for the first time.

They came from North America, Europe, Asia, the Middle East, and emerging robotics hubs that had barely appeared on the map a year earlier. It was the first time the full landscape of humanoids existed in one place.

There was an energy in the room that is difficult to describe without sounding overstated. It was not excitement for its own sake. It was recognition. The builders who had never been in the same room before suddenly saw one another's work with full context. The operators who had been searching for credible paths into humanoids discovered practical steps rather than speculation. Researchers who had spent years exploring the building blocks of autonomy saw how their work fit into deployment. Investors gained a more accurate understanding of timing, constraints, and adoption. The Summit made visible the shape of the field in a way that no one had experienced before.

The program reflected that intent. The focus stayed on real work. The content centered on perception, control, dexterity, learning, generalization, safety, system architecture, and early deployment. The people on stage were builders presenting the details behind their decisions. They spoke about failure modes, uncertainty, tradeoffs, and the realities of deploying autonomous physical systems in environments designed for people. This level of visibility was not common anywhere else. It created shared understanding. It allowed the field to learn from itself.

The program itself was designed with precision. Every session had a purpose and every topic was chosen to move the field forward. Keynotes grounded the audience in the realities shaping deployment. Individual talks gave builders the room to explain design choices, tradeoffs, and performance constraints. Panels were structured to extract lessons that cut across hardware, autonomy, safety, and operations. Live demonstrations were placed carefully to show real capability without drifting into spectacle. A successful gait transition, a controlled manipulation sequence, or an adaptive perception pipeline carried more weight than any polished narrative. The point was to make the work concrete. Claims needed to be visible. Progress needed to be verifiable.

This level of curation created a common reference frame. It helped founders communicate trajectory, helped operators evaluate readiness, and helped investors separate momentum from noise. It made conversations sharper because the groundwork had already been laid on stage.

Transparency was a core principle from the beginning. The Summit encouraged candor without asking builders to reveal sensitive details. Performance numbers were presented with the conditions that produced them. Limitations were discussed openly because the field could not afford false signals. Hidden marketing was not allowed to masquerade as neutral insight. When conversations reached areas that required confidentiality, they moved to side rooms or private meetings rather than being diluted on stage. This clarity built trust. It allowed people to engage honestly, and it set a tone that shaped every edition that followed.

What emerged was something more powerful than a typical conference dynamic. The Summit created a sense of coherence. Builders found collaborators. Operators identified credible partners. Researchers saw which problems were closest to deployment. Investors understood where the deepest technical work was happening. Policymakers gained visibility into the questions that mattered. The conversations did not end when sessions ended. They continued in hallways, side rooms, and private meetings. People stayed longer than they planned. They told me the Summit felt like a necessary correction for a field that had been building without a shared center.

As the first edition concluded, it became clear that the Summit had not only reflected the momentum of the field. It had accelerated it. Builders began referencing one another's work more directly. Operators began forming deployment partnerships. Investors began shaping theses grounded in real constraints. The Summit provided something the field had lacked until that moment. It gave humanoids a foundation that could support the next phase.

The months that followed confirmed this shift. The Summit was no longer just an idea or an event. It had become a structural component of the ecosystem. People used it to orient their work. They prepared for it. They timed announcements around it. They shaped road maps in alignment with it. The Summit had turned into the place where the global field met itself. It became the center of gravity that humanoids had needed since the beginning.

This was the purpose the Summit was built to serve. It created alignment where it had not existed. It created visibility where it had been missing. It created context where the field had been operating without it. It allowed humanoids, embodied AI, and spatial intelligence to move forward not as scattered efforts, but as a coordinated movement.

Scaling the Summit into a Global Ecosystem Engine

The success of the first Silicon Valley edition created a shift in the field that was larger than the event itself. The Summit had become the place where humanoids gained shape and clarity. It gave builders, operators, researchers, and investors a shared reference point. It transformed early signals into structured understanding. And it showed that the category carried enough momentum, depth, and seriousness to sustain its own platform. But the moment the inaugural Summit ended, it became clear that this was only the beginning. If the event was going to serve as real infrastructure, it needed to grow into something that matched the scale, diversity, and global nature of the work ahead.

The idea of expansion did not come from ambition alone. It came from the needs of the ecosystem. Humanoids were not emerging from a single region. They were taking shape across continents, across disciplines, and across industries. Builders in the United States were working on autonomy, manipulation, and system integration. Researchers in Europe were advancing balance, perception, and actuation. Suppliers in Asia were developing motors, sensors, power systems, and manufacturing pathways. Operators in dozens of countries were considering early trials. No single geography held the full picture. The field could not mature without a global platform that united these perspectives into a coherent whole.

The request for expansion began almost immediately. Founders from Europe asked when the Summit would cross the Atlantic. Investors with international portfolios asked for a format that would connect their networks across regions. Researchers who had been working in isolation wanted visibility among peers who shared their discipline. Operators in Asia asked when the event would come closer to their centers of activity. The field was giving a clear signal. It needed a platform with global reach.

Responding to that signal required a deliberate choice. The question was no longer whether the Summit should expand, but how to do so without diluting the qualities that made it valuable in the first place.

Scaling the Summit required a careful approach. Growth for its own sake would weaken the impact. The event needed to remain focused, rigorous, and grounded in real work. It needed to preserve the clarity and intimacy that defined the first edition. The challenge was to expand the platform while keeping the structure intentional. The model had to travel without losing its essence.

Europe The expansion to London became the natural next step. Europe had a deep history in robotics, engineering, and embedded systems. It had long-standing research institutions, strong industrial robotics companies, and a growing interest in humanoids. It also had a perspective shaped by regulated environments, precision engineering, and long-term deployment planning. Bringing the Summit to London created a bridge between European talent and the emerging momentum in the United States. It allowed the field to stretch across continents without losing coherence.

From the start, the London edition carried the same principles as the Silicon Valley gathering. It centered on the real work. It brought founders, operators, researchers, and investors into direct conversation. It focused on technical depth rather than buzzwords. It maintained the same discipline around curation, structure, and tone. The effect was immediate. The European community saw itself reflected in a format that understood the complexity of humanoids. American builders met their European counterparts. Asian suppliers found a central meeting point for collaboration. The Summit became a reference point not only for progress, but for alignment across regions.

The expansion did more than widen geographic reach. It revealed how global the category already was. Builders from Germany, France, the United Kingdom, Scandinavia, Spain, and Central Europe brought approaches shaped by decades of multidisciplinary engineering. Investors with specialized theses found founders whose work matched their long-term horizon. Operators saw a clearer path for adoption. Researchers found a platform where academic depth could connect to commercial readiness. The field gained dimension. It gained texture. It gained a global voice.

Asia The decision to expand to Tokyo followed naturally. Asia was already one of the most active regions in humanoids. Japan held a unique place in robotics history. It had decades of experience with humanoid research, advanced hardware capability, and cultural familiarity with robotics as part of everyday life. China and South Korea were building rapidly in adjacent areas. Manufacturers across the region were developing components, subsystems, and full-stack capabilities that directly supported humanoid deployment. Asia was too central to remain a distant satellite in the global conversation. It needed a direct role.

Bringing the Summit to Tokyo completed the early structure of the global ecosystem. It created a three-region model that matched where humanoids were being built, tested, and deployed. It provided rhythm to the field's progress. Each edition informed the next. Insights from Silicon Valley shaped the conversations in London. Lessons from London sharpened the priorities for Tokyo. The global arc allowed the field to learn in cycles rather than in isolated steps. It allowed people to cross-pollinate ideas, compare approaches, and understand how different regional strengths fit into a cohesive movement.

From Global Summit to Global Platform

The expansion into London and Tokyo followed a straightforward model. Both were direct extensions of the Summit itself. London allowed the platform to establish a presence in Europe, where robotics research, engineering talent, and industrial expertise were concentrated across a relatively compact ecosystem. Tokyo extended the Summit into Asia, one of the most active regions in humanoids, and created a three-region structure that reflected where much of the world's work in robotics, autonomy, and intelligent systems was taking place.

As these editions matured, a different type of opportunity began to emerge. The requests were no longer coming only from attendees, exhibitors, sponsors, or speakers. Governments, economic development agencies, robotics associations, and ecosystem leaders began asking whether Humanoids Summit could establish a presence within their own regions. These conversations were not centered on hosting a one-time event. They reflected a broader desire to strengthen local humanoid ecosystems while remaining connected to the global movement forming around the category.

The interest came from a diverse set of regions. South Korea was investing heavily in robotics, manufacturing, and physical AI initiatives. Singapore was continuing to position itself as a center for innovation, deep technology, and startup development. Brazil represented one of the most significant technology ecosystems in Latin America. The Gulf region was making substantial investments in artificial intelligence, automation, and future industries. Mexico offered a unique combination of manufacturing capability and growing technology adoption. Similar discussions emerged across parts of Europe where local organizations wanted stronger connections to the broader global robotics ecosystem.

These conversations revealed an important limitation. While direct expansion into London and Tokyo made sense, the same model could not scale indefinitely. A centralized approach would eventually constrain growth. At the same time, allowing independent regional events to emerge without coordination would create fragmentation, inconsistent quality, and disconnected ecosystems.

The solution became the Humanoids Summit Global Network.

I designed the program around a simple principle: Regional ecosystems should be able to develop their own identity while remaining connected to a larger global framework. Rather than treating expansion as a collection of unrelated conferences, the program created a structure through which qualified local partners could organize regional editions while maintaining alignment with the broader Summit platform.

This approach allowed local ownership without sacrificing global coherence. Regional partners understood their markets, institutions, industries, and policy environments better than any centralized organization could. At the same time, participants benefited from access to a global network of founders, researchers, investors, operators, suppliers, and policymakers already connected through the Summit.

The objective was not expansion for its own sake. The objective was to create a network of connected nodes that could strengthen the ecosystem as a whole. Knowledge could move more efficiently between regions. Builders could access new markets and partnerships. Investors could gain visibility into emerging innovation centers. Policymakers could learn from deployment experiences occurring elsewhere. Operators could compare approaches across industries and geographies. Each region could contribute its strengths while benefiting from participation in a larger system.

What began as a Summit was gradually evolving into a platform. The Humanoids Summit Global Network represented another layer of infrastructure, one designed to reduce fragmentation, improve coordination, and help humanoids mature as a truly global field. In many ways, it reflected the same philosophy that shaped the Summit from the beginning. The goal was not simply to gather people. The goal was to create the structures that allow a movement to scale.

As the Summit grew, its structure matured. The event became more than a gathering. It became an engine that powered the ecosystem. Builders used the Summit to reveal progress in autonomy, manipulation, perception, and system integration. Operators used it to understand readiness, constraints, and deployment pathways. Investors used it to refine their theses and identify where the real opportunities were forming. Researchers used it to ground their work in real conditions. Policymakers used it to understand what embodied intelligence required at the national level.

The Summit created a rhythm that the entire field began to rely on. It provided a consistent point of calibration. It reset the conversation at regular intervals.

As visibility grew, the Summit also began drawing senior engineers, product leaders, and executives from major technology companies who wanted a clear view of what was emerging. Many attended quietly, observing the work with a level of focus that signaled how seriously they viewed the category. They were not there to explore a trend. They were there because they recognized that humanoids would influence the direction of AI, hardware, and automation. The Summit became their reference point for understanding where the field was headed and how quickly it was moving.

Major media organizations began treating the Summit as the most credible window into the field. Each edition drew reporters, documentary crews, and live broadcast teams from around the world. They were not chasing spectacle. They were documenting a shift in robotics history that was becoming visible for the first time. Their coverage helped translate the technical depth of the work into broader understanding. It gave the general public a clearer view of what humanoids were, how they were being built, and why the movement mattered. The Summit became the place where the world first witnessed the category take real shape.

It separated progress from noise. It gave people a place to see what mattered, what was emerging, and what still needed work. The consistency of the platform became as important as the content itself. The field needed that structure to move forward with discipline rather than haste.

With scale came new responsibilities. The Summit needed to maintain depth, even as the audience grew. It needed to preserve clarity, even as new players entered the field. And it needed to protect the tone that defined its early editions, the grounded seriousness that kept the work at the center. The guiding principle remained unchanged. The Summit was a platform for real progress, a stage for the people building the future of humanoids. It was not a marketing conference. It was the place where the field aligned around the steps needed to move from prototypes to reliable deployment.

The global model also strengthened the rest of the ecosystem. It amplified the work of the venture fund by giving founders a platform that matched their ambition. It created a natural conduit for standards discussions, since the most influential voices across safety, interoperability, and testing were now meeting in coordinated cycles. It supported policy efforts by giving governments a clear view of where the field was headed. It strengthened supplier networks by giving them direct access to builders and operators. It helped researchers form collaborations that translated into practical impact.

The most visible effect was the way the field began to speak with one voice. Before the Summit, humanoids felt like an emerging concept stretched across continents. After the Summit expanded globally, the field began to see itself as a unified discipline. The conversations carried more nuance. The community understood where deployment was realistic, where it was premature, and where more foundational work was required. People understood one another's constraints. They understood the pace of progress. They understood how to move forward together.

The Summit also elevated the perception of humanoids in a way that mattered to the general public. By presenting the field through rigorous, grounded, and thoughtful discussions, the Summit helped widen understanding. It showed humanoids not as speculative machines, but as disciplined engineering efforts. It presented the real work rather than the sensational headlines. It helped build acceptance by framing the category through clarity, purpose, and real-world progress.

As the Summit continued to mature, its impact became more visible in the relationships it created. Founders launched partnerships that originated in hallway conversations. Operators began pilots with companies they discovered at the event. Investors formed syndicates that aligned with the Summit's perspective. Researchers joined development efforts with companies they met through shared sessions. Standards groups shaped early recommendations informed by Summit discussions. Policymakers referenced Summit insights when designing early frameworks. The Summit had become more than the sum of its parts. It had become the engine of coordination for the global humanoid ecosystem.

The impact of the Summit became measurable within months. Partnerships formed that had not been possible when the field was scattered. Operators who attended out of curiosity returned ready to design pilot programs. Investors recalibrated their frameworks around more realistic cost curves, reliability thresholds, and time horizons. Several companies traced key customer conversations, strategic hires, or technical collaborations back to meetings that began at the Summit. For ALM Ventures, the effect was immediate. What had been a noisy and fragmented deal landscape became a consistent flow of credible, technically rigorous opportunities. The Summit clarified who was ready for the next phase, who was still proving fundamentals, and where capital could accelerate progress rather than subsidize avoidable missteps.

The Summit became the mechanism through which the field learned to see itself as a unified discipline. It became the place where progress turned into coherence. It became the platform that carried the ecosystem from early momentum to sustained movement. With that foundation in place, the next phase began to take shape.

Looking back, the expansion was not simply growth. It was infrastructure. It was the architecture required for a global field to function. It created the flow of information that allowed the category to mature. It gave structure to a movement that was too important to grow without one. And it provided a stable foundation for the next chapters in humanoids, embodied intelligence, and spatial intelligence.

That level of alignment created the conditions for the next layer the field required: structure.

Standards and Safety: The Structure Beneath Deployment

As global alignment improved, the need for structure became impossible to ignore. Early builders were making steady progress, and interest from operators was shifting from curiosity to evaluation. Systems were moving beyond isolated prototypes into environments where performance, reliability, and safety would be tested under more realistic conditions. Despite this progress, there was no shared framework for how these systems should be evaluated, compared, or trusted.

Humanoids operate in environments designed for people. They move through spaces that are not simplified or constrained, and they interact with humans, tools, and objects that were never designed with machines in mind. This creates a different set of expectations compared to traditional robotics. Safety extends beyond physical interaction and includes behavior, predictability, and recovery under uncertainty. A system that performs well in isolation but fails unpredictably in a live environment cannot be considered reliable, regardless of its technical sophistication.

Without shared standards, each stakeholder was forced to define expectations independently. Operators evaluating early systems lacked consistent criteria. One deployment might be considered acceptable under one internal framework and rejected under another. Regulators had no clear baseline to reference. Builders spent time translating the same concepts across different audiences. Investors struggled to distinguish durable progress from isolated demonstrations. The absence of structure did not prevent progress, but it made that progress harder to interpret and harder to trust.

Early engagement with standards groups became necessary. It created the foundation for a shared language across builders, operators, researchers, and institutions. It allowed performance to be discussed in consistent terms rather than subjective interpretations. It also began to define how safety could be approached in a field where systems are expected to operate continuously in human environments rather than controlled industrial settings.

These efforts reduced ambiguity at a moment when ambiguity could slow the field. In a system as complex as humanoids, clarity compounds. When expectations are aligned, builders focus on solving real problems instead of re-explaining them. Operators make decisions with more confidence. Capital moves with better signal, and progress becomes easier to measure and compare.

Standards also shifted how progress was evaluated. Early demonstrations focused on whether a system could complete a task. As discussions matured, the emphasis moved toward consistency over time, reliability under variation, behavior when conditions change, and recovery from failure. These dimensions matter more for real deployment than isolated success cases.

The Missing Layer in Deployment

As this structure began to take shape, a limitation became clear. Standards could define how systems should be evaluated, including safety thresholds, testing methods, and performance expectations, but they did not define how systems should be managed once they were deployed in real environments. This gap became increasingly visible as the field moved closer to operation.

Standards also began to influence how decisions are made in real environments. In a warehouse setting, an operator evaluating a humanoid system does not need to interpret raw technical claims.

They need to understand whether the system can complete a defined task within known conditions, how often intervention is required, and how performance holds over time. A shared framework allows those questions to be answered in comparable terms across different systems.

The same applies in manufacturing environments, where tolerance for variation is low and consistency matters more than peak performance. A system that meets defined evaluation criteria can be integrated into existing workflows with greater confidence. Without that structure, each deployment becomes a separate negotiation of assumptions. With it, decisions become repeatable, and trust becomes transferable across sites and operators.

From Standards to Deployment: The Missing Layer

As systems moved from controlled evaluation toward real-world deployment, the focus shifted from performance at a single point in time to behavior over time. The challenge was no longer limited to measuring capability, but extended to understanding how systems evolve once they are operating continuously in environments where conditions change and interaction is constant.

Behavior over Time These questions did not sit within the scope of technical standards. Standards could define how to test a system and what acceptable performance looks like under specific conditions, but they did not define how that system should be governed after deployment.

This gap became more visible as the field moved closer to real-world operation. When engineers, operators, suppliers, investors, and institutions were brought into the same discussions, alignment at a high level often diverged in practice. A system that appeared robust in controlled testing could behave differently under real operating conditions. A capability demonstrated once did not always translate into repeatable performance. A metric that appeared strong in isolation could mask limitations that became visible only over time.

These differences reflected perspective rather than disagreement. Engineers focused on technical capability. Operators focused on reliability and continuity. Investors focused on trajectory and scalability. Policymakers focused on safety and accountability. Without a shared way to translate between these perspectives, alignment remained partial.

Misaligned Perspectives This created friction that extended beyond engineering. Builders were required to align expectations across multiple audiences, often explaining the same system repeatedly in different terms. This slowed progress and introduced unnecessary complexity. At the same time, the absence of shared interpretation distorted signal. Some systems appeared more advanced than they were because they produced compelling demonstrations, while others made meaningful progress in reliability and integration without producing visible moments that could be easily communicated.

Over time, the field began to converge around more practical evaluation criteria. The focus shifted toward consistency across conditions, predictable behavior under variation, and structured recovery from failure. This improved the quality of discussion and made it easier to compare systems in meaningful ways.

Even with this progress, a core question remained unresolved. Standards clarified how systems should be evaluated, but they did not define how systems should be governed once deployed. This gap sat between engineering and liability and was not fully addressed by any existing framework.

Deployment Governance: The HOMES Framework

Through ongoing work with founders, operators, and early deployments, this gap became increasingly visible. Even as standards improved, there was no shared model for how humanoids should be introduced, managed, and maintained once they left controlled environments. Each deployment was approached independently, with each operator defining its own expectations and each company developing its own internal processes. There was no consistent structure connecting these efforts.

This created coordination risk. Systems might perform well, but uncertainty around responsibility, updates, and incident handling made deployment slower and more difficult than necessary. The limitation was not technical capability, but the absence of a shared operational model.

To explore one approach addressing this, I developed a working framework called HOMES, short for Humanoid Operating Model for Environment Safety.

Introducing HOMES HOMES is not a technical standard. It does not define how systems should be built or how performance should be measured. It is an operational governance model intended to guide how humanoids are deployed, managed, and maintained in environments where humans and machines interact continuously.

The framework reflects the reality that a deployed system is not static. It evolves over time, receives updates, encounters new conditions, and interacts with people in ways that cannot always be predicted in advance. Governance must account for this continuity.

HOMES introduces a structured approach to deployment as an ongoing process rather than a one-time validation. A humanoid system deployed in a warehouse or manufacturing setting does not operate under static conditions. Tasks evolve, environments change, and systems receive updates that affect behavior over time. Governance needs to account for this continuity.

In practical terms, this means treating deployment as a managed lifecycle. A system is introduced with clearly defined operating conditions, monitored as it performs under real workloads, and continuously evaluated as those conditions shift. Identity and traceability ensure that behavior can be linked back to specific system states and updates. Operational boundaries define where and how the system is expected to perform, reducing ambiguity when conditions change. Human oversight is explicitly defined, including when intervention is required and how control is transferred. Incident logging creates a shared record of failure modes and recovery patterns, allowing lessons from one deployment to inform others.

This structure does not eliminate failure, but it makes failure visible, contained, and learnable. Over time, this is what allows systems to move from supervised operation to more autonomous behavior without introducing unmanaged risk.

Without a shared model, these elements remain inconsistent. Each company defines them differently, and each operator interprets them differently. This does not prevent deployment, but it increases friction and reduces trust. It also limits the ability of the ecosystem to learn collectively.

As deployment approaches real-world scale, these questions intersect with existing legal and regulatory frameworks. Product liability, workplace safety, and operational accountability become immediate considerations. Early discussions with legal and industry stakeholders have explored how governance models such as HOMES could align with existing legal structures.

The objective is not to introduce unnecessary constraints. It is to reduce uncertainty at the point where systems operate in real environments. Clear expectations make deployment more predictable. Defined responsibilities make risk manageable. Structured learning allows progress to compound.

Trust is established through consistent operation over time, not isolated demonstrations. As deployment became more tangible, it also became clear that governance alone was not sufficient. The ecosystem needed coordination at a broader level.

Regional Anchoring: Bay Area Robotics Association (BARA)

I created the Bay Area Robotics Association to give the region a steady, neutral place to work together between major moments of alignment. It was formed in direct response to sustained demand from founders, operators, investors, and corporate teams who needed a member-driven platform focused on execution rather than visibility. The Summit is where we gather at scale. Standards work is where technical baselines are set. Policy is where public-path friction is cleared. BARA is the day-to-day table where the people responsible for delivery meet, compare notes, and keep moving when there is no stage and no press. It is not another show. It is a regional utility that runs year round.

A Working Layer

Before BARA, the region lacked a regular, neutral, membership-based forum that combined capital access, buyer linkage, and practical execution support for robotics and embodied AI. Conversations existed, but they were fragmented. Capital moved, but often without enough operational context. Pilots started, but lessons did not reliably circulate. No standing structure existed to connect startups, corporates, suppliers, and investors in a way that was both routine and grounded in real work. BARA was built to close that gap.

The Bay Area has a density of founders, corporate teams, labs, suppliers, and capital that you can feel in a single block. That density is a strength, but it also creates noise. Signals scatter across private chats, small dinners, and one-off introductions. Important updates get trapped inside calendars and inboxes. A supplier that could shorten a lead time never meets the integrator who needs that capability. A corporate group searching for a pilot site has no line of sight to the local team with the right safety posture. A graduating lab team leaves the area because they did not know two nearby companies would have hired them immediately. None of these misses are dramatic in isolation. Over time, they drain energy and momentum.

Hidden Inefficiencies

BARA exists to close those small gaps before they become large delays. It is a place where a founder can hear what buyers are asking for this quarter, where an operations lead can understand what a lab validated last month, where a supplier can see how their part fits into a credible bill of materials, and where investors and corporate development teams can engage startups with a clearer view of real deployment timelines. Connecting capital to credible execution is not an accessory. It is one of the association's core functions. The Bay Area moves quickly. A standing forum is the only way to keep the right people in the same conversation without waiting months for the next large event.

From the outset, BARA was designed with external anchors as part of its structure, not as an afterthought. While the Bay Area serves as the operational center, the association launched with deliberate linkages to other major robotics regions, including the US East Coast, Tokyo, and Shanghai. The intent is practical rather than symbolic. Robotics supply chains, capital formation, and deployment patterns already span regions. A regional forum that ignores that reality leaves its members to discover constraints late and individually. BARA shortens that distance by giving members early visibility into how work, capital, and expectations move across regions.

For members, this matters in concrete ways. A startup gains earlier signal on where demand is forming and how readiness is evaluated outside its home market. A supplier understands how capacity decisions in one region affect timelines in another. A corporate group can benchmark its deployment assumptions against similar programs operating under different conditions. Investors see how capital stacks, partnerships, and go-to-market paths vary across regions before those differences surface in negotiations. None of this requires global expansion from day one. It requires shared awareness, so decisions made locally do not collide with realities discovered later.

Calling the organization the Bay Area Robotics Association is intentional. The Bay Area name carries weight not because it claims centrality, but because it signals openness, interoperability, and translation. This region has long served as a place where technologies are shaped to operate across environments, industries, and markets. BARA reflects that role. It is a regional table with a global line of sight. Members do not join to "go global." They join to avoid building blind spots into systems that will eventually operate beyond one geography.

Neutral means no single company sets the agenda. Useful means every gathering produces information that helps someone take a next step. The format stays simple. Short briefings with only the facts needed to act. Context from people who are doing the work. Time carved out for direct follow-up so the right pairs can continue without spectacle. There is no product theater. No attempt to settle debates about architectures. The focus stays on the details that unlock progress: site rules that actually work, insurance language that will bind, test methods buyers trust, logistics constraints that cannot be waved away, and lead times that are real.

That approach sets the tone. People do not show up to grandstand. They show up to compare notes. When a team shares good news, they also share the tradeoffs. When a pilot report looks clean, supervision hours and incident handling are part of the discussion. When something fails, the specifics are covered plainly so the next team does not repeat the same path. This is not a court and not a press conference. It is a work session.

No Theater

BARA is designed to be boring in the right ways. It runs on member dues and transparent sponsorships, so cadence does not depend on the news cycle. The schedule is regular enough to plan around and light enough not to exhaust participants. Programming follows what the region is actually doing, not what would make a headline. Governance stays simple. Roles are clear. No products are certified. No companies are blessed. No grants are allocated. The job is to hold the space and keep it reliable so the people building and deploying robotics can find each other and keep moving.

The cadence is predictable by design. BARA operates on a bimonthly rhythm that members can plan around, with regular briefings, working sessions, and targeted introductions. Matchmaking is structured and qualified. Introductions are made when there is a clear reason to engage, whether that is a pilot opportunity, a sourcing question, a capital discussion, or a hiring need. The goal is not volume. It is relevance. Members should leave with fewer open loops, not more names.

Beyond meetings, the association provides a small set of shared services that reduce friction across the region. A members-only directory makes it easier to find the right counterpart without broadcast outreach. Closed-door sessions allow sensitive topics to be discussed without press or posturing. Short written summaries capture decisions and constraints, so lessons persist beyond the room. These are simple tools, but they compound. They turn informal knowledge into usable context and help members move faster without cutting corners.

Clear boundaries make this work possible. Sessions are not sales pitches. Private road maps are not demanded. Nothing shared in confidence is published. Time is spent translating across roles so no one talks past anyone else. A lab update is framed for a site manager. A buyer requirement is unpacked so a founder can map it to a plan.

An insurance clause becomes a checklist an operator can use on the floor. Meetings end with shared understanding rather than polite confusion.

The same boundaries apply to disagreement. Competing approaches are treated as data, not drama. Teams explain the conditions where their approach performs, the cost drivers that matter, and the common failure modes. Buyers hear enough to align expectations with reality rather than promises.

BARA is not an island. It feeds and is fed by the other layers. Lessons from a working session inform standards discussions the following month. A buyer update becomes a short case that a Center of Excellence can format and share. A method refined at a national lab shows up in a site visit where operators see how to apply it. A theme that keeps surfacing becomes a Summit discussion with leaders who can move budgets. A policy friction point becomes a short memo that helps clear it. The loop matters more than any single event.

A founder arrives convinced a customer wants a broad capability. They leave with a narrower task the customer will certify and pay for, along with the insurance clause required to bind coverage. A supplier expects lumpy demand for a year. They leave with visibility into two programs that justify a modest capacity investment. A corporate team considering a pilot hears what supervision actually costs and calibrates accordingly. None of these moments make news. All of them move the field forward.

Robotics is at a stage where a bad assumption can cost a quarter. A misunderstanding about site rules can stall procurement. An optimistic read of a lead time can derail a schedule. An unstated expectation around incident handling can stop a deal cold. Fewer unforced errors matter more than more declarations. BARA reduces those errors by making the obvious explicit and the tacit shareable.

Timing matters for talent as well. The region produces people with the exact skills needed to keep pilots honest and deployments running. Those paths should be visible. BARA helps companies find people faster and helps people land where they can contribute immediately. That loop compounds.

A forum that depends on hype is fragile. BARA is designed to be durable. Dues are predictable. Sponsorship is transparent and tied to shared programming, not access. The calendar is steady. Formats repeat. When the news is loud, the pace stays measured. When the news is quiet, nothing is invented. The work continues.

If this works, people will talk about what they accomplished, not the forum they used. The deployments and operations will be the proof. The forum's role is to make those outcomes more likely, month by month, with cleaner execution and fewer preventable delays.

International Humanoids Day: A Cultural Anchor for the Field

The need for a cultural anchor could not be addressed through messaging alone. It required something repeatable, visible, and grounded in real work, something that could hold meaning over time and extend beyond the core industry.

That anchor became International Humanoids Day.

What Took Shape

The intent was simple. If humanoids were going to enter human environments and remain there, the field needed a recurring moment to reflect on why they existed, how they were being used, and whether progress was aligning with real human needs. A single day could not carry the entire cultural burden of an emerging technology, but it could establish rhythm. It could create a shared reference point and signal a long-term commitment to responsibility rather than spectacle.

I introduced International Humanoids Day alongside the inaugural Humanoids Summit and set for December 12. The date was chosen deliberately. Practically, it coincided with the first global gathering focused exclusively on humanoids, grounding the day in real work rather than abstraction. Symbolically, 12/12 carries a deliberate symmetry. Humanoids exist alongside humans, not apart from them. They emerge from the pairing of artificial intelligence and embodiment. They require software and hardware to advance in lockstep. Progress in this field depends on balance across these paired domains. Overweight one side and the system fails. The mirrored date reflects that structure: two sides evolving together, not competing for dominance. Humanoids sit at the intersection of intelligence and embodiment, software and physics, cognition and action. The mirrored date reflected that duality.

12/12 Symmetry

Progress in this field depends on balance, not dominance of one side over the other.

The framing was equally intentional. The guiding principle was humanoids for humanity.

That phrase was not meant as a slogan. It was a constraint. It set expectations about where value should accrue and how success should be measured. Humanoids were positioned as systems designed to extend human capability and dignity, not to replace people wholesale or chase novelty. In practical terms, this meant focusing on environments where physical strain, safety risk, or labor scarcity constrained outcomes. Warehouses. Manufacturing floors. Logistics hubs. Infrastructure maintenance. Places where consistency and endurance mattered more than theatrics.

Early examples made this tangible. At one logistics facility, operators watched a humanoid perform repetitive pallet handling under supervision. The task itself was unremarkable. What mattered was what followed. When a minor obstruction caused a fault, the system paused, signaled for assistance, and resumed safely once cleared. For engineers, this was a routine recovery sequence. For operators, it was the first time they saw failure handled predictably rather than catastrophically. The conversation shifted.

Shift in Perception

The question moved from "Will this replace workers" to "How many injuries does this prevent over a year."

Another moment came from a manufacturing pilot where technicians were initially skeptical. They had seen automation cycles before. Systems arrived with promises, struggled under edge cases, and quietly disappeared. This time, the demonstration focused not on peak performance, but on maintenance. The humanoid was intentionally stopped mid-task. A technician walked through inspection, reset, and redeployment. The message was clear. This was not a black box. It was a tool that could be understood, serviced, and improved. Several of those technicians later enrolled in training programs to specialize in system oversight. The technology did not eliminate roles. It created new ones.

These moments mattered because they changed posture. Fear softened into curiosity. Resistance turned into questions. Not because anyone argued abstractly about jobs or productivity, but because people saw systems behaving in ways that respected human presence and control.

International Humanoids Day was designed to surface and repeat those moments. Not as isolated success stories, but as a pattern the field could hold itself accountable to year after year.

This is how trust compounds.

Trust over Time Trust does not emerge from announcements. It comes from visible safeguards, predictable behavior, and clear recovery when things go wrong. It comes from transparency about conditions. What was demonstrated. Under what constraints. With what supervision. International Humanoids Day made those disclosures part of the cultural expectation. Not optional. Not performative. Normal.

Over time, this approach created space for broader participation. Students attended open sessions and asked about training paths. Small manufacturers explored whether a single system could extend operating hours without exhausting staff. Warehouse managers compared notes on downtime and recovery rather than marketing claims. These conversations did not happen because the technology was perfect. They happened because the framing invited realism.

Practical Framing The day was also designed to scale globally. Humanoids were never going to be confined to a single region. Manufacturing expertise, supply chains, research depth, and deployment contexts were distributed across continents. A shared observance allowed different regions to participate on equal footing, even when local priorities differed. A factory in Asia could highlight reliability improvements. A European operator could discuss safety protocols. A North American logistics hub could report productivity gains alongside reduced injury rates. The point was not uniformity. It was comparability.

From the beginning, the intent was for International Humanoids Day to mature beyond an industry moment and into a globally recognized observance. Early discussions are underway to explore formal recognition through United Nations–affiliated cultural and scientific heritage frameworks, following the path taken by other international days that began as community-led initiatives before achieving institutional acknowledgment. The goal is not symbolism for its own sake, but permanence. A durable annual reference point that outlives any single organization, company, or technology cycle.

Other international days provide a useful comparison. World Environment Day. International Women's Day. World Space Day. None of these began as fully institutionalized observances. They started as focal points created by communities that understood the need for shared reflection before broader recognition followed. International Humanoids Day is following the same trajectory, grounded first in real participation and only then seeking formal continuity.

Path to Recognition As humanoids move from early deployment into more common use, the importance of a fixed cultural anchor becomes clearer. Technologies that reshape daily life are rarely remembered for their first demonstrations alone. They are remembered for the moments when society learned how to relate to them. When expectations stabilized. When fear gave way to familiarity. International Humanoids Day is intended to become that marker for embodied intelligence. Not the day humanoids appeared, but the day the field committed publicly to introducing them with care.

This commitment also provided a way to address concerns directly rather than dismiss them. Workforce displacement fears were not waved away. They were met with examples. In one distribution center, a humanoid system reduced overnight strain by handling repetitive lifting, allowing teams to rotate into higher-skill supervisory roles. Injury rates dropped. Overtime stabilized. Retention improved. The system did not replace a workforce. It reshaped it.

Learning curves became visible as well. Early deployments were clumsy. Recovery times were long. Supervision was constant. Over successive cycles, downtime shrank. Training materials improved. Operators began anticipating failure modes before they occurred. These improvements did not come from breakthroughs. They came from iteration. From feedback loops. From the discipline of returning each year and reporting honestly.

International Humanoids Day gave those iterations a public cadence. It made improvement legible.

Over time, the day can expand into distributed observances. Facility tours where visitors see complete task cycles rather than highlights. Technical workshops where students learn calibration and safety checks. Regional programs that bring demonstrations to communities far from major technology hubs. The first year focused on establishing principles and rhythm. The future is about widening participation without diluting intent.

The long-term impact is not measured in headlines. It is measured in consistency.

What Actually Matters Fewer injuries. Shorter recovery times. More technicians trained. More sites moving from pilot to steady operation. More small suppliers achieving reliable demand. These are the signals that matter.

At scale, the result is a time dividend. When routine physical effort costs less, human attention can shift. First toward safety and training. Then toward quality and care. Over years, this compounds into healthier workplaces, better-maintained infrastructure, and more resilient supply chains. The change is gradual. It is not dramatic. And it lasts.

International Humanoids Day was created to support that arc. It does not dictate outcomes. It holds space for them. It gives the field a moment each year to pause, measure, and recommit. That is what cultural anchoring looks like when it is built, not declared.

With a stable cultural reference in place, the ecosystem gained something essential. Not momentum, but direction. Not hype, but memory. A way to ensure that as humanoids become more capable and more common, the reasons they exist remain visible and shared.

That foundation made the next challenge unavoidable. Once systems are understood and accepted, society begins asking not just how they work, but how they fit. Addressing that question requires more than culture. It requires governance, policy, and long-term stewardship. That is where the work turns next.

Policy Framework: The Humanoids Act Proposal

As humanoids moved from controlled pilots toward real deployment, the need for a clear policy framework became unavoidable. Standards were beginning to define how systems should be evaluated. Governance models were emerging to guide deployment. What remained was a broader structure that could align research, safety, manufacturing, and adoption at a national level.

Policy, in this context, was not about restriction. It was about coordination. It created clarity for builders to move forward, for operators to commit, and for institutions to support the field without slowing it down.

As humanoids began to appear in real pilots, a consistent pattern emerged. Builders were advancing quickly. Buyers were interested but cautious. Standards work was progressing, but not always connected to deployment. This was not failure. It was a coordination gap. Left unaddressed, it would fragment progress, slow adoption, and increase the cost of responsible deployment.

That is the context in which I drafted a proposal for a US Humanoids Act. Not as finished legislation, but as a framework for productive federal engagement. It is written in clear, accessible language so policymakers can approach humanoids constructively, grounded in how systems are actually built, tested, bought, and operated.

I have seen how fast-moving fields drift when structure lags progress. Aviation, energy, and semiconductors benefited from early frameworks that aligned research, standards, and industry without locking in outcomes. Embodied AI now sits at a similar inflection point.

The question is not whether humanoids will advance, but whether the surrounding structure keeps pace in a way that preserves safety, credibility, and long-term capacity.

The proposal starts from a simple premise. Humanoids fit human environments. That is their advantage. They move through spaces built for people and interact with tools already in use. When a technology aligns with the world as it exists, the appropriate policy response is coordination that removes friction and lowers risk while safety and workforce planning move in step with deployment.

Fit to Human Environments

The timing reflects a balance. It avoids writing rules for problems that do not exist, while also avoiding leaving assumptions to be set elsewhere and absorbed later. The framework does not push mandates, pick technical winners, or lock in architectures. It focuses on pilots, test methods, early procurement, and training. These tools create predictable cycles, turn scattered experiments into an organized learning process, and support the transition from demonstration to bounded, repeatable tasks.

Humanoids are still early. That makes scope discipline essential. A focused framework can define what is reasonable now, what should be measured, and how incidents are handled. It keeps attention on tasks that match current capabilities and enforces realism around supervision, intervention, and uptime. This does not require a large new bureaucracy. It requires a small program office, a limited set of testbeds, and alignment with existing standards.

The core intent is coordination, not control. The goal is to reduce uncertainty at the point where systems meet real environments. In practice, this means a small program office with a defined mandate. It convenes industry, researchers, and standards bodies on a regular cadence. It publishes a short annual road map grounded in real deployments. It tracks progress using task-level metrics rather than announcements. It removes routine obstacles that slow pilots, from site rules and insurance ambiguity to mismatched expectations around supervision and incident reporting.

Safety and standards are integrated from the start. The framework builds on existing work across standards bodies and national labs, favoring task-level certification because it is faster to validate and easier to understand. A buyer can interpret what it means to certify a specific task within defined conditions. When pilots reference the same methods, comparisons hold across sites and vendors, and trust follows from consistent measurement.

Standards in Practice

The proposal also introduces a US Center of Excellence with a narrow, connective role. It links standards development, testbeds, buyer requirements, and vendor capabilities. It publishes practical guidance that sites can reuse, including task templates, supervision checklists, incident reporting formats, and insurance frameworks tied to certified tasks. It does not centralize control. It centralizes know-how so lessons move efficiently across regions.

This framework depends on credible methods. It emphasizes support for national labs and leading research centers to develop test methods, datasets, and protocols that make evaluation repeatable. These institutions can run reference trials, publish baselines, and validate results across independent sites. When those baselines are visible and stable, buyers know what they are getting, and builders know how they will be measured.

Testbeds remain grounded in real work. Companies bring their systems into measured environments that reflect where they intend to deploy. Logistics corridors, factory settings, and controlled public facilities provide bounded conditions where performance can be observed over time. Trials are time-boxed, with clear entry and exit criteria. Results are documented through practical case studies that show performance, failure modes, intervention rates, and uptime.

Early procurement bridges the gap between testing and use. Agencies can run limited pilots aligned with current capabilities, such as materials movement in controlled environments or structured inspection tasks. Each pilot follows a repeatable process for certification, insurance, reporting, and review. Over time, this creates a body of reference deployments that reduce uncertainty and support broader adoption.

Policy, in this context, is not separate from the ecosystem. It is part of the structure that allows it to function.

Regional Nodes: How the Ecosystem Became Global

As the ecosystem expanded, a pattern began to take shape that went beyond individual initiatives or regions. What initially looked like geographic growth started to reveal itself as something more structured. The field was not spreading randomly. It was organizing itself into distinct centers of capability, each contributing a different layer to the advancement of humanoids.

At first, this was not intentional. It emerged through observation. Different regions were solving different parts of the problem, often without direct coordination. Over time, those differences became clearer, and more importantly, they became complementary. What one region lacked, another had already developed. What appeared fragmented at a distance began to look like a system when viewed up close.

This shift required a change in perspective. Instead of viewing progress through the lens of individual companies or technologies, it became necessary to understand how capabilities were distributed across geographies. The question was no longer who was ahead, but how the pieces fit together across regions.

Complementary Regional Strengths

The United States pushed forward on artificial intelligence, capital formation, and early platform thinking. It became the center of gravity for software intelligence and venture-backed scaling. Japan and South Korea continued to lead in precision robotics, components, and reliability, building on decades of engineering discipline and manufacturing excellence. China developed unmatched strength in manufacturing scale and supply chain execution, enabling rapid iteration and cost efficiency. Europe focused on regulatory frameworks, safety structures, and standardization, shaping how systems could be deployed responsibly.

Each region advanced along its own path, shaped by its economic structure, talent base, and industrial history. These differences were not temporary. They were structural. They reflected long-standing strengths that could not be easily replicated elsewhere.

Humanoids could not scale from any one of these regions alone. The systems required to bring them into real environments depended on all of these layers working together. Intelligence without manufacturing would stall. Manufacturing without intelligence would limit capability. Deployment without standards would introduce risk. Capital without coordination would fragment progress. The field required alignment across regions, not dominance by a single one.

The Need for Cross-Regional Alignment

As this became clear, the role of the ecosystem began to shift. It was no longer enough to build platforms, host events, or support companies within a single geography. The challenge became connecting these regional strengths in a way that allowed knowledge, capital, and capability to move across borders without friction.

This was not just about collaboration. It was about reducing structural inefficiencies. Without alignment, each region would move forward with its own assumptions, its own benchmarks, and its own interpretations of progress. That would lead to duplication of effort, incompatible systems, and delayed adoption.

Alignment did not mean centralization. It meant creating enough connection points so that progress in one region could inform decisions in another. It meant ensuring that the field evolved as a system rather than as a set of disconnected efforts.

One of the earliest and most visible of those connection points was the Humanoids Summit.

The Humanoids Summit naturally evolved into one of the first points of connection. With editions across Silicon Valley, London, and Tokyo, it created a shared space where regional perspectives could meet directly. Builders from Asia could engage with investors from the United States. European regulators could interact with operators and engineers working on real deployments. Conversations that would have remained isolated within regions became part of a broader dialogue.

The value of this was not visibility. It was context. Participants could see where they stood relative to others. They could compare approaches, challenge assumptions, and identify gaps. This created a level of clarity that was difficult to achieve within any single region.

Over time, the Summit helped establish a common frame of reference. Progress was no longer interpreted through isolated signals. It was understood through direct comparison and shared discussion. This reduced ambiguity and accelerated decision-making across the ecosystem.

These connections gradually extended beyond individual regions and began forming a broader global network.

BARA extended this logic at a more localized level, anchoring activity in Silicon Valley while maintaining openness to global participation. It created continuity between gatherings, allowing relationships formed at the Summit to persist and evolve. What began as a regional association became a node within a larger network, connected to other clusters, institutions, and industry groups across different parts of the world.

Other regions began to form their own anchors, whether through associations, research centers, or industry partnerships. These were not copies of each other. Each reflected local priorities and strengths. But they shared a common function: maintaining continuity and facilitating exchange.

Over time, these efforts started to form a network of interconnected nodes rather than isolated hubs. Each node retained its local strengths, but none operated in isolation. Information flowed between them. Lessons learned in one region influenced deployment strategies in another. Standards discussions began to reflect a more global perspective. Investment decisions incorporated insights from multiple geographies rather than a single market view.

As these connections strengthened, the risk of fragmentation began to decline.

This structure reduced one of the field's biggest risks: fragmentation. Without coordination, regions would have progressed in parallel, developing incompatible systems, conflicting standards, and uneven levels of maturity. That path would have slowed adoption and increased integration costs.

Fragmentation does not always appear immediately. It builds quietly. Different assumptions lead to different implementations. Different implementations lead to incompatibilities. By the time those differences surface, they are often difficult to reverse.

By creating points of connection early, the ecosystem allowed the field to move with greater coherence. Differences still existed, but they were visible and could be addressed. The goal was not to eliminate variation, but to prevent divergence from becoming a barrier to scale.

The benefits extended beyond coordination alone.

It also changed how progress was evaluated. Instead of viewing advancements as isolated breakthroughs, they could be understood within a broader context. A manufacturing improvement in one region could be seen alongside advances in perception models in another. A policy framework developed in Europe could inform deployment strategies in the United States.

The field began to develop a shared sense of direction. This did not come from a central authority. It emerged from repeated interaction, comparison, and alignment across nodes. Over time, this created a more stable trajectory for the industry.

Geography did not disappear in this process. It became more important. Each region continued to operate within its own constraints and priorities. But instead of acting as boundaries, those differences became inputs into a larger system. The goal was not uniformity. It was alignment.

What emerged was more than a collection of individual initiatives.

This shift marked an important transition. The ecosystem was no longer defined by the initiatives themselves, but by how those initiatives connected across regions. What had started as a set of platforms, organizations, and programs became a coordinated structure capable of supporting a global movement.

The strength of that structure came from its distribution. No single node controlled it. No single region defined it. Instead, it relied on the interaction between nodes, each contributing its part to the overall system.

In the early stages, this coordination remains incomplete. Gaps still exist. Alignment continues to evolve. But the direction is clear. Humanoids are not developing within isolated markets. They are emerging through a network of regions, each shaping the outcome in different ways.

Understanding that structure changes how the field is approached. It shifts the focus from local optimization to global coordination.

It reframes competition as part of a broader system. And it provides a foundation for scaling humanoids in a way that reflects the complexity of the world they are designed to operate in.

Putting the Pieces Together—Infrastructure as Strategy

I did not build separate projects. I built parts of one system. The fund, the Summit, standards work, a cultural anchor, policy, and a regional forum all serve the same goal. They make it easier for real companies to build real humanoids that work in human spaces. Each piece solves a different kind of friction. Together they turn a scattered field into a movement with momentum.

One System, Not Many

Early on, I saw a simple pattern. Capital without shared context wastes time. Events without follow up fade. Standards without field data drift. Policy without practical examples becomes talk. Regional meetups without clear links to national work get inward. If these parts stay separate, deal flow stays shallow, pilots stall, and buyers lose patience. The answer was to wire them together so information and trust could move in both directions.

That wiring started with company building. ALM Ventures sets the spine. It gives founders a partner who understands the constraints of embodied systems and the timelines of hardware. The first check is not enough. The value is in helping teams avoid the dead ends you can see from the pattern of deployments and pilots across the ecosystem. That is why the first investment was not a company.

It was the Humanoids Summit. I needed a place where the field could look at itself, align, and reduce waste.

The Summit was never meant to be a stage. It is a working room. I kept curation tight so builders, buyers, researchers, investors, and policymakers could speak plainly. Closed sessions when needed. No recordings. Clear attribution rules. People came ready to share what was true. The effect was twofold. It showed the field to itself, and it gave people enough signal to make better calls when they left. A founder learned how a buyer actually certifies a task. An insurer explained the clause that would bind coverage. A standards lead saw which test methods would get used. Those details travel far after the lights are off.

Going global early was a deliberate choice. London and Tokyo were not expansion for its own sake. They prevented geographic silos. They kept language aligned. They created early partners for standards and policy work. The mirror effect and the platform effect reinforced each other. The room reflected what was real. The visibility gave good work the credibility it needed outside the room.

That field truth mattered most when standards entered the picture. The conversations happening in the Summit did not stay in the room. They shaped how standards bodies, testing groups, and applied research labs approached the work. The focus stayed practical. Start with tasks. Make methods specific. Keep the door open for different architectures as long as behavior meets the test.

When those same methods show up in real deployments, the loop is working. When insurers and operators can rely on them, the loop becomes stronger. Standards are not paperwork. They are how trust scales.

Trust also had to extend beyond the builders themselves.

Public understanding had to evolve alongside the technical work. I created International Humanoids Day just before the first Summit and announced it during the event. December 12 is a simple anchor. It gives the public a moment to look in. It gives the field a reason to explain what it is doing in accessible terms. Cultural anchors matter because they hold the story steady. They also push teams to account for progress and safety in words the public understands. This is not marketing. It is a habit that keeps the technology grounded in human impact.

Policy entered as quiet scaffolding, not as declaration. The Humanoids Act proposal is a draft, not a bill. I wrote it to give Congress and agencies a practical starting point if they choose to act. It frames humanoids as a national competitiveness priority. It favors pilots, test methods, early procurement, and workforce plans over speeches. It calls for a small program office, time-boxed testbeds, a Center of Excellence to move know-how, and support for national labs and university centers to build shared artifacts and methods. The tone is calm. The aim is productive.

All of this becomes real at the regional level. BARA is where the system shows up day to day. In the Bay Area, density is both a feature and a risk. Without a steady forum, signals scatter. With BARA, briefings, working sessions, and site visits turn private knowledge into common sense. A team adjusts a plan based on a buyer's real process. A supplier commits to capacity after seeing predictable demand tied to certified tasks. A lab shows a method that operators can use the next day. We keep the table neutral and useful. That alone saves quarters of time.

Over time, the loop compounds. Each piece feeds the others. The fund learns where teams stall and where they speed up. That learning shapes Summit topics and connects the right people. Summit sessions produce case studies and methods that standards groups can reference. Standards clarity keeps the policy draft concrete and modest. Policy clarity makes it easier for BARA to help a buyer and a vendor try a narrow task with a clean stop rule. Results from that pilot feed back into the fund's view of what to back next. This is what I mean by infrastructure as strategy. You do not push any one piece too hard. You keep the loop healthy.

How the Pieces Reinforce Each Other

The benefits are small at first and obvious in hindsight. Time from first meeting to a qualified conversation shortens. Pilots produce better write-ups because methods are common and supervision models are clear. Insurance binds faster. Suppliers see steadier demand and invest in quality. Hiring lines up with what operators actually need. These are not headlines. They are the conditions that let a field grow up.

What this approach demanded most was restraint. Infrastructure work is slower than company building and quieter than platform launches. It does not reward urgency the way venture culture is trained to expect. There are no immediate usage curves to point to and no single metric that captures progress. Instead, the work shows up as fewer misunderstandings, shorter cycles, and cleaner handoffs between actors who would otherwise never align. Those gains are real, but they only become obvious after the fact.

The incentives rarely line up. Capital wants velocity. Founders want momentum. Corporates want certainty. Policymakers want clarity without risk. Infrastructure offers none of those quickly. It requires holding space before demand is visible and doing the unglamorous work of alignment before scale is possible. Most ecosystems avoid this phase or try to outsource it to chance. They assume standards will emerge on their own, that trust will follow adoption, and that policy can catch up later. In fields where software dominates, that sometimes works. In embodied systems that move through human environments, it does not.

Why This Field Is Different

That is why this had to be built deliberately and early. Humanoids surface every weak seam in the system at once. Hardware, autonomy, safety, insurance, labor, procurement, and public trust collide in the same deployment. If even one layer is missing, progress slows or reverses. The temptation is to push harder on the visible layer and hope the rest sorts itself out. I chose the opposite. I worked on the invisible layers first so the visible work would hold when it arrived.

The sequencing followed from that choice. Starting with capital established seriousness and staying power. Creating the Summit next was not about attention but shared context. Standards came early so pilots would not fragment into incompatible claims. Cultural anchoring preceded policy so public understanding could mature alongside the technology rather than react to it. Policy followed once there were enough concrete examples to keep it practical and restrained. The regional forum came last to make sure none of this drifted between cycles.

Changing that order would have broken the loop. Policy without field truth would have hardened the wrong assumptions. Standards without buyers in the room would have optimized for theory rather than use. Events without capital behind them would have produced conversation without commitment. A regional forum without national and global links would have turned inward. Each piece depended on the credibility created by the one before it.

This is infrastructure as strategy in practice. Strategy is often framed as market capture or technical advantage. Here, strategy is about reducing systemic waste. It makes it easier for honest work to compound and harder for weak signals to masquerade as progress.

Infrastructure does not replace innovation. It filters it. It raises the floor so good teams spend their time building rather than explaining, renegotiating, or correcting misaligned expectations.

What makes this kind of strategy uncomfortable is that it runs against how progress is usually rewarded. Infrastructure does not announce itself. It does not produce instant validation. It is judged later, often by people who were not present when the decisions were made. In the moment, it can look like hesitation rather than intent, or coordination rather than leadership. That ambiguity is part of the cost. You accept it because the alternative is visible motion without durable direction.

This work also requires absorbing risk that is hard to price. When you build companies, risk is bounded by the company. When you build infrastructure, the risk is systemic. If you sequence incorrectly, you do not just slow one effort. You distort many. If you move too early, you harden assumptions. If you move too late, fragmentation sets in. There is no clean rollback. That reality forces a different kind of discipline. Decisions have to be made with incomplete data but long consequences in mind.

It also requires patience with outcomes that do not map neatly to ownership. Infrastructure creates shared advantage. It benefits competitors as much as collaborators. It improves conditions even for teams you do not back and sites you do not control. That is not a bug. It is the point. The payoff is not exclusivity. The payoff is a field that behaves more rationally, wastes less time, and compounds faster once real demand appears. You trade short-term differentiation for long-term coherence, trusting that strong work will still stand out when the ground beneath it is stable.

This approach also reflects how humanoids will ultimately scale. Adoption will not hinge on a single breakthrough or a single company. It will hinge on whether many teams can deploy safely, repeatedly, and predictably across different environments. That requires shared language, shared methods, and shared expectations long before mass deployment.

The infrastructure is the moat.

It allows variety without chaos and competition without fragmentation. There is discipline in knowing when not to act. Infrastructure work requires saying no to shortcuts that create momentum without durability. It requires resisting pressure to declare victories early or collapse complexity into slogans. It requires patience with slow consensus when fast alignment would fail later. Those are not fashionable traits, but they matter when the goal is not a headline but a field that can sustain itself.

The result is a system that absorbs new entrants without resetting. New founders arrive into clearer expectations. New buyers see comparable claims. New suppliers see demand signals that justify investment. New policymakers see examples that are modest and real. The work becomes cumulative rather than episodic.

The time compression matters. None of this unfolded over decades. It was assembled quickly because the window was narrow. The field was early enough that norms were still fluid, but mature enough that mistakes would have real consequences. Building infrastructure in that window required moving fast without rushing and sequencing without hesitation. That balance is what allowed the pieces to lock together rather than compete for attention.

With the infrastructure in place, the next task is the physical platform itself.

4

The Architecture of a Humanoid

The infrastructure had to be constructed around humanoids before they could move from promise to practice. Capital, convening, standards, policy, and regional coordination were not parallel efforts. They were prerequisites. Without them, progress would have remained fragmented and fragile.

Humanoids are systems where sensing, motion, computation, balance, and human interaction collide in one body, operating continuously in real environments. Every architectural decision carries consequences that ripple outward, affecting safety, reliability, insurance, procurement, and public trust.

The architecture of a humanoid explains why standards had to mature early, why pilots mattered more than demos, why buyers moved cautiously, and why building the surrounding ecosystem was not optional.

Before a humanoid ever reaches for an object or takes a step, it inherits a set of constraints that no other class of machine carries all at once. It must stand and move in spaces built for human bodies, not for rails or fixed paths. It must operate for long periods without rest, close to people who do not think in terms of safety envelopes or test conditions. And it must do so while remaining predictable, legible, and interruptible.

This is why humanoids progressed differently from software and from traditional industrial robots. Software can fail quietly and recover invisibly. Industrial robots succeed by isolation. Humanoids have neither luxury. They are exposed systems. Every choice in their design shows up in public, often immediately, and often in ways that are difficult to reverse.

To make sense of a humanoid as a system, we return to architectural responsibilities. I think of them as pillars not because they stand alone, but because each one carries a different kind of failure. When one weakens, the entire machine feels less trustworthy, even if everything else appears to work.

None of these pillars exist in isolation. They lean on each other constantly. A sensing error becomes a balance problem. A computation delay turns into a manipulation mistake. An actuation limitation shows up as social discomfort. Architecture is the discipline of designing these dependencies so that stress does not collapse the system.

The Five Pillars

The five pillars emerged from watching systems fail in ways not connected to a single discipline. The real cause for robotics hesitation often lived in the space between a timing mismatch, a thermal limit, and a calibration drift that no one flagged. Over time, the failures formed into patterns that pointed to where reality and expectation refused to connect.

Together, these pillars define whether a humanoid can operate reliably in the real world. A weakness in any one of them eventually surfaces somewhere else in the system. Strong sensing cannot compensate for poor actuation. Advanced manipulation cannot overcome unstable balance. More compute cannot solve unreliable sensing. Each pillar contributes a different capability, but their real importance lies in how they interact under real-world conditions.

Sensing: Where Reality Pushes Back First

When people talk about humanoids, they usually start with intelligence, focusing on smarter models, better reasoning, or more general behavior. I learned fairly early that this framing misses the problem that shows up first once robots leave tightly controlled settings and are asked to perform in front of real people, even for limited periods of time.

In practice, most humanoid failures do not begin with bad decisions. They begin with perception that appears solid during preparation and testing, but becomes fragile once conditions start to shift, even within the span of a single extended demonstration or repeated live session.

I saw this pattern repeat across projects, teams, and public demos. Systems could be carefully planned and technically impressive, yet still slow down, hesitate, or behave unpredictably as conditions changed. The issue was rarely that the robot chose the wrong action. More often, it was responding to a world that had already moved on by the time the response arrived.

That realization changed how I think about sensing. I stopped treating it as a collection of inputs feeding intelligence and started seeing it as the most fragile contract in the entire system, one that everything else depends on and one that real environments stress almost immediately.

In lab settings or carefully staged demos, perception often looks solved. Lighting is controlled, backgrounds are clean, and objects are placed intentionally. Under those conditions, sensors perform well and confidence builds quickly, making it easy to believe that perception is largely behind us as a problem.

Humanoids, however, are shown in spaces that are shaped by clutter, reflections, heat, dust, vibration, and constant improvisation. Once robots are exposed to these environments, even briefly, it becomes clear that perception is never clean or stable, regardless of how advanced the models appear on paper.

Light changes across the day, shadows stretch and collapse, and reflective surfaces interfere with depth estimation in subtle ways. Warehouses combine natural light, fluorescent flicker, and localized glare, while manufacturing environments add dust, heat, and vibration.

What stood out was not simply that errors occurred, but how they surfaced during operation.

The robot rarely failed outright. Instead, it slowed down, hesitated, or behaved conservatively in ways that looked like intelligence limitations from the outside but were rooted in perception uncertainty underneath.

The pattern was consistent enough that I began cataloging the failure modes. As systems became more complex, timing emerged as a constraint alongside accuracy. Humanoids rely on multiple sensing streams at once, including vision, depth, force, inertial signals, and proprioception, each arriving at different rates with different latencies and noise characteristics. Keeping those signals aligned in time proved far more difficult than fusing them in space.

During longer live demos and repeated sessions, it was common to observe systems reacting to conditions that were already gone. Balance corrections arrived late, hands closed on objects that had shifted, and recovery behaviors triggered after the moment had passed. From the outside, these behaviors resembled control issues, but at the system level, they reflected timing mismatches that intelligence alone could not compensate for.

As models grew heavier and compute pipelines lengthened, this tension increased. Perception stacks competed with planning and control for shared resources, latency budgets tightened, and behaviors that worked during short runs degraded as demonstrations extended or were repeated, leaving systems capable in theory but less responsive in practice.

Degradation also appeared earlier than many teams expected. Sensors accumulated dust, calibration drifted, thermal noise increased, and mounts loosened under vibration. Even across repeated demos and pilot activity, perception systems spent more time operating away from ideal calibration than founders initially assumed.

Operators planned around frequent recalibration and tightly controlled conditions, only to discover that even early exposure strained those assumptions. When perception quality degrades faster than maintenance or intervention allows, sensing stops being a purely technical concern and becomes an operational constraint that no improvement in models can immediately resolve.

What proved more valuable than incremental model gains were unglamorous foundations. Consistent sensor supply chains, calibration practices that worked across vendors, data pipelines that captured failure cases rather than just successes, and maintenance procedures that did not require specialized expertise all mattered more than advertised accuracy.

This was the first place where I saw the gap between what individual teams could learn and what the field needed to know. Ambiguity intensified these pressures as robots moved closer to real tasks. Humans are comfortable operating with partial information, while humanoids are not. Objects are occluded, tools are misplaced, and materials are arranged in ways that violate training assumptions. Temporary workarounds make sense to people but confuse machines, forcing perception systems to operate with incomplete context even in early pilot scenarios.

In these situations, perception has to balance confidence against caution. Too much confidence increases safety risk, while excessive caution undermines usefulness. That balance is never purely technical. It is shaped by environment, application, and tolerance for risk, even at the demonstration and pilot stages.

Different contexts tolerate different errors. A logistics-style demonstration allows for different behavior than a tightly constrained manufacturing scenario, while public-facing environments demand conservative behavior and social trust. Sensing architecture has to encode these tradeoffs explicitly, because hiding them behind aggregate accuracy metrics only postpones the consequences.

As exposure increased, a consistent operational reality emerged. When perception failed, founders needed to understand why at a system level, with enough clarity to diagnose lighting effects, occlusion, drift, latency, or misclassification without interpreting raw sensor data.

This requirement reshaped how sensing systems had to be built. Diagnostics became essential, logs needed to be actionable, and data had to flow back into retraining pipelines so that failures informed future behavior.

At that point, it became clear that sensing was never going to be solved inside a single robot. It required shared datasets, common failure definitions, cross-company benchmarks, and standards for reporting and remediation, along with forums where teams could speak honestly about what broke without turning it into marketing. That level of coordination only emerges at the ecosystem level.

Sensing also dismantled the myth of full autonomy by showing that perception defines a negotiated boundary between what a robot can handle on its own and when assistance is required. Architectures that acknowledged this reality and supported graceful fallbacks delivered value earlier than those chasing purity.

Actuation and Power: Where Physics Stops Negotiating

If sensing taught me where systems first begin to drift, actuation taught me where optimism runs out even faster.

Early on, I thought of actuation as a problem that disciplined engineering would eventually tame. With the right components, enough testing, and careful integration, motion could be made reliable through iteration. That assumption held as long as systems stayed inside short demonstrations and controlled conditions. It weakened once robots were asked to perform for longer stretches, across repeated sessions, and under environments that refused to stay consistent.

In the beginning, humanoids capable in isolation struggled quietly as physical demands accumulated. The issue was whether movement could be sustained without steadily eroding precision, responsiveness, or confidence as time passed.

The moment intention turns into motion, abstraction disappears. Actuation lives in a world governed by mass, friction, inertia, heat, and wear, and none of those forces care how advanced the software stack is. Even within a single extended demonstration, these effects begin to surface.

Humanoids operate in contact-rich environments where objects vary in weight and compliance, surfaces shift unexpectedly, and human intervention is common. Actuation in those settings has to deliver useful force while remaining forgiving enough to avoid damage or injury, and that balance shifts continuously based on context, accumulated stress, and the state of the system in that moment.

Across repeated demos and early pilot activity, a consistent pattern emerged: Systems that appeared crisp at the beginning of a session would begin to show subtle changes as time passed. Movements became slightly less precise. Recovery behaviors took longer to trigger or complete.

Heat connected all of these observations and was almost always underestimated. Actuation, power electronics, and energy storage all generate heat within a tightly constrained body, and paths for dissipation are limited. What looks acceptable during short, scripted motion becomes a constraint as demonstrations extend or are repeated back-to-back.

It was common to observe gradual slowdowns during longer live sessions without any obvious external trigger. Control margins narrowed, precision slipped slightly, and the system continued functioning while delivering less than it had at the start. Thermal throttling rarely announced itself directly, but its effects were unmistakable once systems were observed over time rather than in a single snapshot.

These behaviors were not signs of missing intelligence or poor planning. They reflected architectural assumptions about duty cycles, cooling, and energy use that had not yet been fully tested under prolonged physical activity. Addressing them required coordination across mechanical layout, power distribution, thermal strategy, and operational expectations, not isolated improvements in software or control logic.

Energy efficiency compounded all of these pressures. Humanoids draw power continuously, and every corrective motion carries a cost. Even within short pilots or extended demos, small inefficiencies accumulated into shorter usable windows and increasing intervention, shaping how systems could realistically be presented or tested.

Watching these dynamics play out shifted my attention away from how impressive a robot looked at first glance and toward how its behavior changed as physical demands accumulated. Endurance, predictability, and consistency began to matter more than expressive motion, even before true deployment was on the table.

Mechanical wear made these lessons clearer with repetition. Bearings loosened, lubrication degraded, and small impacts accumulated stress. In humanoids, wear tended to concentrate in joints compensating for upstream sensing uncertainty or control delay, creating uneven degradation even at this early stage.

Because humanoids carry their wear, changes in dynamics forced control systems to work harder over time, amplifying stress elsewhere in the body. Without sufficient architectural margin, these effects compounded rather than stabilizing, even across limited but repeated use.

Understanding where actuation truly breaks emerged from watching similar patterns repeat across different teams, environments, and appearances, until the same weaknesses surfaced again and again.

Power management reinforced this lesson. Power was never just about capacity. It was about how energy was allocated, prioritized, and protected across the body as demands competed continuously. Sudden bursts from recovery behaviors, sustained draw from manipulation, and constant background consumption from sensing and compute all pulled against each other, even in short operational windows.

When power orchestration was poorly integrated, behavior became brittle. Systems performed well in one mode and degraded in another or functioned reliably early in a session and struggled later. From the outside, these machines appeared inconsistent. At the architectural level, they were expressing coordination failures.

Watching these patterns repeat fundamentally changed how I think about motion. Movement alone does not equal capability. Reliable motion emerges only when physics, economics, safety, and maintenance are treated as a single system rather than separate concerns, even at the demo and pilot stage.

Actuation is where ambition meets gravity, and gravity always wins unless the system is built to respect it.

Compute and Control: Where Time Becomes the Bottleneck

Compute and control revealed something subtler but just as limiting: Intelligence itself was rarely the bottleneck. What mattered was whether decisions arrived in time to be acted on safely and consistently.

Early on, I assumed that as models improved, behavior would naturally follow. Better perception would feed better planning, and better planning would produce better motion. That assumption made sense in isolation, but it weakened once systems were asked to run end-to-end, even for extended demos or repeated pilot sessions.

What became clear was that knowing what to do and doing it quickly enough were fundamentally different problems.

Humanoids operate within narrow temporal margins. Balance adjustments, contact transitions, grasp corrections, and collision avoidance all depend on responses arriving within specific windows. When those windows are missed, the system fails physically, with movements becoming hesitant, corrections overshooting, or recovery behaviors triggering too late or unnecessarily.

During early demos, these effects were easy to overlook. Systems performed well enough in short bursts, and smooth behavior early in a session masked deeper timing fragility. As sessions extended or were repeated back-to-back, latency began to surface in subtle ways. Reactions that initially felt confident became delayed, and behavior that appeared stable started to feel cautious or inconsistent.

This was the consequence of compute pipelines carrying more complexity than they could reliably deliver within the time available.

However, as models grew heavier and system stacks expanded, compute increasingly became a scheduling problem rather than a capability problem. Perception, planning, control, safety monitoring, and logging all competed for finite resources. When priorities were not explicit, degradation showed up in unpredictable ways. Control loops starved intermittently, safety checks lagged, and arbitration between behaviors became inconsistent.

From the outside, these issues often looked like logic bugs or control instability. At the system level, they were timing failures rooted in architectural choices that had not yet been stressed under integrated operation.

One lesson emerged clearly from watching these systems evolve: Predictable behavior within known bounds mattered far more than peak performance or impressive reasoning depth once robots were observed over longer sessions.

Early humanoid systems often pushed toward richer models and longer planning horizons because the behavior looked impressive when it worked. Even modest increases in latency, however, reduced trust quickly once timing inconsistency became visible. Systems that responded reliably, even if conservatively, proved easier to operate and safer to demonstrate.

That reality forced difficult tradeoffs. Smaller models, shorter horizons, and simpler decision layers became more attractive despite feeling like a step backward at first. Over time, those choices made systems easier to reason about and more resilient, even at the demo and pilot stage.

Compute also surfaced as a physical constraint rather than a purely logical one. High-performance processors drew power continuously and generated heat inside already constrained bodies. As thermal limits were approached, throttling introduced additional latency, which in turn narrowed control margins and reduced responsiveness. This feedback loop was quiet but persistent, especially during longer demonstrations.

Unlike cloud systems, humanoids cannot scale compute elastically. Every architectural decision is embodied. Choices about model size, scheduling strategy, and numerical precision propagate directly into power consumption, thermal behavior, and responsiveness, even when systems operate only for hours at a time.

Control exposed another limitation that became increasingly difficult to ignore: Abstractions leaked.

High-level planners often assumed clean state estimates and timely actuation. In practice, sensing noise, calibration drift, latency, and mechanical variation introduced discrepancies that propagated upward through the stack. When control architectures were not designed to absorb those gaps, behavior oscillated or degraded rather than stabilizing.

These effects became more visible when humans entered the loop. People move unpredictably, interrupt tasks, and violate expectations. Control systems had to arbitrate safety, intent, and motion in real time, often under compute pressure. Conservative behavior emerged because timing uncertainty made aggressive responses unsafe.

As these patterns repeated, it became clear that compute and control could not be treated as a thin layer beneath intelligence. They had to be first-class architectural concerns, with explicit guarantees, bounded failure modes, and clearly defined escalation paths.

Compute was defined by time, coordination, and predictability under constraint. Control was never just about algorithms. It was about absorbing uncertainty without amplifying it.

Manipulation and Dexterity: Where Usefulness Becomes Visible

For a long time, manipulation was treated as something that would naturally emerge once the rest of the humanoid stack improved. Better perception told the robot what to grab, better control guided how it moved, and better models decided what came next. Dexterity was assumed to be downstream of intelligence and motion rather than a primary constraint in its own right, an assumption that seemed reasonable when most progress was evaluated in controlled settings.

That view changed once systems were observed interacting with the physical world, even in limited demonstrations and early pilot conditions where environments were carefully prepared and tasks were tightly scoped.

Repeated observation showed that manipulation is where intelligence is constrained, exposed, and often humbled. This is the point where humanoids stop being judged on promise and start being judged on whether they can reliably reduce friction in environments built around human hands and tools, rather than merely occupying space within them. Walking draws attention because it is visible and emotionally resonant, and standing upright signals progress even when little work is being done. Hands do not produce the same reaction, yet they determine whether a humanoid can perform meaningful tasks once it is standing there. A robot that moves convincingly but manipulates unreliably remains peripheral in human environments, regardless of how advanced the rest of the system appears.

The gap between how easy manipulation looks to humans and how difficult it is for machines is wider than it initially appears. Humans pick up tools, orient objects, apply force, and adjust grip continuously without conscious thought, often while distracted or multitasking. Each of these actions depends on tightly integrated feedback loops involving touch, force, prediction, and correction, operating at a level of coordination that current humanoid systems still struggle to reproduce, even in narrow task domains.

Part of this gap comes from how humans acquire dexterity in the first place. Manipulation skills are learned through years of continuous embodied interaction with the world, where mistakes are frequent, feedback is immediate, and adaptation is constant. Humanoid systems, by contrast, are expected to generalize from far more limited exposure, often under constraints that restrict how much physical exploration is possible.

In early demonstrations, manipulation often appeared competent within narrow bounds. Objects were familiar, placements were anticipated, and environments were prepared to minimize variation. Under those conditions, performance looked stable enough to suggest that dexterity was largely solved, or at least close enough that remaining gaps would close quickly.

Once those assumptions loosened even slightly, performance degraded in ways that were immediate, uneven, and difficult to recover from, particularly when variation accumulated rather than appearing all at once.

These breakdowns made it especially difficult to diagnose clear failures. Instead of dropping objects or aborting tasks outright, systems often entered a pattern of hesitation, retry, and incremental adjustment that looked deliberate from the outside. Grips tightened and loosened without converging, end effectors hovered just off contact, and motions slowed as if the system were being cautious, when in reality it was accumulating uncertainty it could not resolve.

Small changes in object orientation, surface friction, placement, or occlusion frequently triggered retries, slowdowns, or the need for intervention. The small margin that existed between success and breakdown revealed how tightly constrained effective manipulation still was, even in scenarios designed to be forgiving.

This narrow margin explains why manipulation has earned its reputation as one of the hardest remaining problems in humanoids, even though the difficulty is often misattributed to the hand itself. The reason manipulation is so often framed as a hand problem is that the hand is where failure becomes visible. When a grasp slips or a tool misaligns, the breakdown appears localized, even though the cause is rarely confined to the end effector alone.

In practice, manipulation exposes how tightly coupled the system really is. Small compromises upstream in perception timing, posture, balance, or control precision eventually surface at the point of contact, where there is no buffer left to absorb them. Focusing narrowly on hand design can therefore be misleading, because it treats the symptom rather than the condition. Dexterity does not emerge from a better hand in isolation, but from the consistency and coordination of everything that leads up to it.

Manipulation is the convergence of multiple uncertainties at the moment of contact with the world. Perception errors feed directly into grasp planning, contact physics introduce variability that models struggle to predict, control latency amplifies small mismatches, and mechanical compliance changes behavior over time. Because manipulation sits at the intersection of all of these forces, it exposes weaknesses that other subsystems can temporarily conceal.

Early efforts to address this often emphasized mechanical sophistication, adding joints, articulation, and degrees of freedom in the hope that human-like form would produce human-like capability. In practice, increased complexity introduced new fragilities, including calibration drift, thermal buildup, wear, and tighter synchronization requirements across sensing and control, which reduced reliability rather than improving it.

Other approaches moved in the opposite direction, simplifying end effectors to gain robustness at the expense of generality. These systems often performed better on narrowly defined tasks, but struggled as soon as conditions deviated from what they were designed to handle.

The role of touch made this limitation especially clear. Vision can estimate where an object is and how it is oriented, but it cannot reliably convey weight, compliance, micro-slip, or contact quality. Without tactile feedback integrated at sufficient bandwidth, manipulation becomes cautious and indirect, with conservative grips, frequent pauses, and behaviors shaped more by uncertainty than by intent.

Tasks requiring subtle force modulation make this especially visible. Turning, inserting, aligning, or fastening objects demands coordination across space, time, and force simultaneously, and even small deviations can lead to failure.

Two-handed manipulation compounds these challenges further. Coordinating both hands introduces dependencies that are difficult to manage in real time, as forces applied by one hand affect alignment and stability in the other. Timing mismatches tend to amplify errors rather than cancel them, and recovery becomes more complex because there are more contact points and more possible failure paths.

This is where the distinction between capability and usefulness becomes unavoidable: A system that can complete a manipulation task occasionally, under supervision, does not create value at scale. Usefulness emerges when manipulation works across variation, without constant adjustment, and without demanding attention, a bar that remains well above what most systems can reliably deliver today.

It also became clear that dexterity does not reside solely in the hand. Posture influences reach and force application, balance affects stability during contact, and timing determines whether corrections improve or worsen outcomes. Many manipulation failures originate upstream and only become visible at the fingertips, making manipulation an unusually honest test of system integration rather than a localized capability.

Anthropomorphism alone does not resolve this challenge. Human-like hands are appealing because the world is designed for human tools, but copying form without matching function creates frustration rather than capability.

Over time, manipulation forced difficult tradeoffs into focus, including generality versus reliability, speed versus safety, and expressiveness versus endurance. These tradeoffs cannot be resolved through better models alone. They involve decisions about acceptable failure rates, intervention thresholds, and economic value that extend beyond the robot itself.

As a result, isolated development stopped working. No single team encounters enough variation to understand manipulation deeply, and failures are too context-dependent to generalize from limited exposure. Meaningful progress required shared learning across tasks, environments, and system designs, along with common ways of describing contact breakdowns and recovery behavior.

Manipulation made it clear that reliable partial capability often creates more value than brittle completeness. Systems that acknowledged their limits, escalated when uncertainty exceeded those limits, and recovered predictably proved more useful than systems pursuing general dexterity but failing unpredictably under variation, particularly in tasks where small errors carry outsized consequences.

Today, manipulation remains the most demanding test of humanoid viability. Walking allows humanoids to enter human spaces, but manipulation determines whether they can participate meaningfully in them, and progress in this area will depend less on isolated breakthroughs than on alignment across sensing, actuation, compute, control, task design, and operations.

Balance and Locomotion: Where Stability Is Negotiated

Balance was the first capability that made humanoids feel real to people. The moment a machine could stand upright, shift its weight, and take steps without external support, it crossed a psychological threshold. Walking became shorthand for progress, as if movement alone implied readiness for real work.

That reaction is understandable. Humans are bipedal. Balance is so deeply ingrained that it disappears from conscious thought. Standing upright feels effortless, even though it is one of the most complex things the human body does. We forget that balance is not a static state, but a continuous process of sensing, prediction, correction, and adaptation.

Watching humanoids struggle with balance forced me to revisit that assumption. What feels natural to humans is anything but simple to replicate, even in controlled settings. For decades, balance was one of the clearest reminders that robots were not built for the same world we were.

Historically, balance was difficult because robots lacked the sensing fidelity, compute speed, and control bandwidth to react quickly enough. Small delays cascaded into instability. Minor errors amplified into falls. The problem was physical and unforgiving.

Balance hasn't been solved, but it's more manageable. Improvements in sensing, faster control loops, better actuation, and more robust estimation have narrowed the gap enough that upright posture and walking no longer feel impossible. Even in short demos today, humanoids can stand, step, and recover in ways that would have seemed out of reach not long ago.

That progress, however, creates a new illusion. Because robots can now walk convincingly for brief periods, balance is often treated as a completed milestone rather than an ongoing constraint.

In controlled environments, locomotion can look deceptively stable. Floors are flat. Surfaces are known. Paths are planned. Under those conditions, balance appears smooth and repeatable, reinforcing the idea that the hard part is behind us.

Human environments introduce variability everywhere. Floors slope subtly. Materials change friction without warning. Cables, thresholds, and debris appear unexpectedly. Even during short demonstrations, these details matter. Balance margins that look generous in simulation narrow quickly once variability enters the picture.

Instability rarely announced itself dramatically. Falls were uncommon in managed demos. Instead, instability showed up as effort. The robot worked harder to remain upright. Corrections became more frequent. Motion slowed slightly.

Balance was the convergence of the pillars. Balance exposed how tightly coupled the entire stack really was. Perception noise shifted estimates of body position. Actuation delays slowed corrective forces. Compute latency reduced responsiveness. Locomotion issues were rarely isolated problems. They were integration problems expressing themselves physically.

During extended demos or repeated sessions, this coupling became more visible. Systems that appeared stable at the beginning required increasing correction as time passed. Energy use rose. Heat accumulated. Control margins narrowed.

Locomotion amplified these effects further. Walking is energetically expensive, even over short distances. Starting, stopping, turning, and sidestepping introduce additional demands that simple forward motion does not capture. When robots carried objects or adjusted posture mid-task, balance became even more sensitive to timing and coordination.

If bipedal locomotion is this demanding, why insist on it at all? From a purely task-optimized perspective, bipedal form is often unnecessary. Wheels are more efficient. Fixed bases are more stable. Specialized platforms outperform humanoids at many individual tasks. That has always been true.

Bipedal form instead offers compatibility. Human spaces are built around human proportions, human reach, and human movement patterns. Stairs, thresholds, doorways, shelves, and work surfaces assume an upright body that can step, pivot, and reposition itself without reconfiguration of the environment.

Balance and locomotion are, therefore, not about access. A humanoid that can move through human spaces without modification unlocks flexibility that specialized machines do not, even if it pays an efficiency penalty to do so.

That tradeoff makes balance a system-level decision about where effort is spent and where constraints are accepted.

What became clear over time is that balance is less about perfect walking and more about recovery. Most of the real work happens between ideal steps. Small slips. Minor misplacements. Slight delays. The system's ability to absorb those disturbances without escalating them matters far more than how graceful its gait looks at rest. Near-falls were more revealing than falls. Hesitation before stepping. Overcorrection after contact. Brief freezes while the system reassesses.

Human presence raised the bar further. When humanoids moved near people, balance became a matter of trust as much as physics. Even minor stumbles altered perception, regardless of whether harm occurred. Stability became a social requirement layered on top of a technical one.

This forced conservative behavior earlier than many teams expected. Wider stances, slower steps, and cautious transitions were not signs of immaturity. They were signals that systems were prioritizing safety and predictability over spectacle, even at the demo and pilot stage.

Over time, it became clear that balance and locomotion could not be evaluated in isolation. Foot design, joint compliance, sensing fidelity, control timing, and surface interaction all influenced outcomes.

Balance and locomotion did not mature because robots learned to walk. They matured because systems learned to recover, yield safely, and remain predictable under imperfect conditions, even when that meant moving more slowly than audiences initially expected.

Integration: Why Humanoids Are Systems, Not Products

None of these constraints behave independently in the real world. Sensing does not fail on its own. Actuation does not degrade in isolation. Compute delays do not remain contained inside software.

Balance problems are rarely just balance problems, and manipulation failures almost always trace back to decisions made elsewhere in the system. What looks like a local weakness is usually a failure of integration.

Products are expected to have clear boundaries, stable interfaces, and well-defined responsibilities. If something fails, it should be possible to identify the component, replace it, and restore functionality. Humanoids do not behave this way. Their failures propagate across layers, accumulate over time, and often only become visible when multiple small compromises align in the wrong order.

The system does not announce where it broke, because it never broke in one place.

Each improvement revealed new constraints downstream. Early on, it was tempting to believe that each pillar could be advanced independently, and that maturity would come from stacking improvements on top of one another. Better sensors would clean up perception. Stronger actuators would handle physical interaction. Faster compute would resolve timing. More data would improve manipulation. Each of these assumptions made sense in isolation.

In practice, improvement in one area often exposed fragility in another. Sharper perception increased sensitivity to latency. Stronger actuation amplified thermal and power constraints. More sophisticated models placed greater strain on scheduling and control. Attempts to improve manipulation revealed weaknesses in posture, balance, and tactile feedback that had been easy to ignore before. Progress did not move in a straight line because the system itself was not linear.

Failures proved that progress in isolation creates new dependencies. What mattered most was not how advanced any single capability appeared, but how well the system absorbed uncertainty when those capabilities interacted.

Humanoids operate in environments that refuse to stay fixed, and the moment a system touches the world, assumptions are tested simultaneously across sensing, motion, timing, and control. When those tests failed, the failures were always integration problems. The gap was integration under constraint.

The consequence was predictable across independent teams. A locomotion team might advance balance and stability only to discover that their improvements exposed weaknesses in perception that had been masked by careful demo staging. A manipulation team might build high-precision hands, only to find that those hands' precise positioning requirements created new control challenges the team had not anticipated. A perception team might create robust vision models but discover that the latency they introduced broke the timing margins that control systems relied on.

Each group was naturally optimized for what it controlled. Locomotion teams focused on strength and durability. Model teams focused on accuracy and generalization. Control teams focused on stability and responsiveness. Operators focused on reliability and safety. All of these perspectives were valid, but none of them were sufficient on their own. The moment one team improved its area, it created a new problem for someone else down the line. No one could see the full picture.

This is where the ecosystem stopped being optional. Progress required shared definitions of failure, because founders needed to agree on what it meant for a humanoid to be "working" in the first place. It required common benchmarks that reflected variability rather than ideal conditions, because success in clean settings did not translate to messy ones. It required data that included breakdowns, retries, and recovery, rather than only successful executions.

The limiting factor was that these contributions were not part of a coordinated system. The moment when coordination became possible, everything accelerated. The pieces did not magically fit together. Locomotion platforms improved enough to support more complex manipulation strategies. Perception models became flexible enough to handle broader variations. Actuators reached performance levels that could support full-body coordination. Compute modules advanced to the point where more sophisticated AI could run onboard.

These shifts revealed the outline of what a more connected field could become. Integration explains how humanoids function as systems.

Failure as the Default State

Integration explains how humanoids function as systems, but it does not capture how those systems behave once they are exposed to time, variability, and real human environments, where performance is shaped less by isolated capability and more by how consistently the system absorbs change.

In early conversations around humanoids, failure is often framed as something exceptional, where a system works until it does not, or a demonstration succeeds or breaks. That framing does not hold once systems leave controlled settings and begin operating in real environments, because behavior does not move cleanly between success and failure. Instead, systems operate through continuous partial failure, adjusting and compensating in real time as conditions shift, with reliability defined not by the absence of failure but by how well the system recovers without escalating the situation or requiring intervention.

This becomes clear as soon as systems move beyond structured environments where variability is minimized. In laboratory settings and staged demonstrations, lighting is stable, objects are known, and interactions follow expected patterns, allowing behavior to appear consistent and repeatable. Under those conditions, it is easy to conclude that reliability has been achieved, because the system is not being asked to absorb meaningful uncertainty.

That perception changes once the same systems are exposed to environments shaped by human behavior rather than controlled inputs. Objects are moved without notice, tools are substituted, surfaces vary in subtle ways, and tasks evolve without explicit communication. People interrupt, improvise, and adjust continuously, often without recognizing that they are introducing variability. Each individual change may be small from a technical standpoint, but together they create continuous pressure on the system, forcing it to operate outside the assumptions under which it was validated.

Under these conditions, systems rarely fail abruptly. Instead, they degrade, with movements becoming slower, grasp attempts requiring adjustment, and balance corrections appearing more frequently. What looks like hesitation or lack of confidence is often the system managing uncertainty it cannot fully resolve, redistributing risk across sensing, actuation, and control in order to maintain stability.

Unlike software systems, where failure can often be contained within layers of abstraction, humanoids express failure through physical behavior that is immediately visible. A delay appears as hesitation, a perception error becomes a misplaced action, and a control issue shows up as overcorrection or instability. Even when a task is ultimately completed, the path taken to reach that outcome influences how the system is perceived, because humans interpret motion, timing, and consistency directly.

This makes failure not only a technical condition but a behavioral signal. Small inconsistencies in movement or timing are quickly read as uncertainty, even when the system is functioning within acceptable limits. In repeated demonstrations and early pilots, systems rarely collapse under pressure but instead enter cycles of hesitation, retry, and adjustment, where grasps are refined incrementally, steps are taken cautiously, and actions unfold with visible effort. These patterns reflect the system negotiating uncertainty across multiple subsystems rather than executing clean, isolated decisions.

Reliability, in this context, emerges from how the system behaves while these imperfections are present. Systems that degrade gracefully maintain predictable behavior as conditions shift, slowing down, simplifying actions, and preserving stability even when precision is reduced. From the outside, they may appear conservative, but they remain understandable and safe. Systems that lack this property respond differently, with small deviations triggering disproportionate reactions, corrections overshooting, and recovery introducing new errors that compound rather than stabilize, leading to inconsistent behavior and increased need for intervention.

These differences are rooted in architectural assumptions. Systems designed around clean inputs and stable conditions tend to break under real-world stress, while systems designed to absorb uncertainty extend their usable range even when individual components remain imperfect. The ability to manage imperfection across subsystems becomes more important than optimizing any single component in isolation.

As systems operate over longer durations, additional effects begin to surface. Sensors drift, thermal conditions change, calibration shifts, and small inconsistencies accumulate, causing behavior that appeared stable over a short interval to diverge when repeated over time. A single successful run provides limited insight into how a system will behave across multiple consecutive runs, where consistency becomes the primary measure of reliability rather than peak performance.

Human presence introduces another layer of constraint, because people interact with systems in ways that are not predictable or structured. They step into paths, move objects mid-task, and expect the system to adapt without explicit instruction, while also interpreting behavior through the lens of human expectations. Movements are expected to be legible, actions are expected to make sense as they unfold, and hesitation or awkwardness can reduce confidence even when the system remains technically correct.

As a result, evaluation extends beyond whether a task is completed to how it is completed, including timing, smoothness, predictability, and the clarity of the system's behavior to the people around it. A task that succeeds after visible retries may be judged as unreliable, while a slower but consistent execution may build trust more effectively.

This changes how autonomy is understood in practice. Systems are often expected to resolve uncertainty internally, but many environments require ongoing interaction between the system and the people around it, whether through explicit assistance or implicit adaptation. Systems that acknowledge this boundary and provide clear points of escalation tend to deliver value earlier than those attempting to eliminate it entirely, because they remain stable under conditions where full autonomy would introduce risk.

Progress, in this setting, is measured less by the introduction of new capabilities and more by how consistently those capabilities are expressed across variation. A system that performs reliably under imperfect conditions, recovers without escalation, and remains understandable to the people around it contributes more to real-world adoption than a system that performs impressively in isolated demonstrations but degrades under repetition or variability.

Failure, in practice, becomes part of the operating condition rather than a binary outcome, shaping how humanoids are designed, evaluated, and introduced into real environments. That shift in perspective leads directly to the next constraint, where operating alongside people requires safety to be embedded into the architecture itself rather than treated as an external layer.

Safety as an Architectural Constraint

As humanoids move closer to people, the conditions described earlier begin to take on a different weight, because failure is no longer evaluated only in terms of performance, but in terms of safety, predictability, and trust. What appears manageable in isolation becomes unacceptable when systems operate in shared environments, where behavior affects not only the task but the people around it.

Safety in this context cannot be treated as an external layer added after capability is achieved. It is embedded in the architecture of the system, shaping how sensing, actuation, control, and interaction are designed from the beginning. Every decision about how a robot moves, how it interprets uncertainty, and how it responds to unexpected conditions carries safety implications that cannot be separated from core functionality.

Traditional industrial robotics approached safety through separation. Robots were placed behind barriers, their workspaces clearly defined, and human interaction limited or controlled. Under those conditions, safety was achieved by preventing contact, and performance could be optimized within a known and bounded environment. Humanoids operate under a different assumption, where coexistence is required and interaction is expected, even in early pilots and limited deployments.

This shift changes the problem fundamentally. When robots and people share the same space, safety becomes a property of behavior rather than a property of containment. The system must remain predictable, interruptible, and bounded in its actions, even as it responds to variability in the environment.

Actuation plays a central role in this. The forces that enable useful work are the same forces that can cause harm if not carefully controlled. Decisions about torque limits, compliance, and motion profiles directly affect how safe a system is in practice. A movement that is efficient in isolation may be unacceptable in a shared environment if it lacks sufficient margin for unexpected contact or interruption.

Control systems must also reflect this constraint. Timing is not only about performance but about ensuring that responses occur within windows that preserve stability and safety. Delays in perception or control can lead to overcorrection, hesitation, or unintended contact, all of which affect how the system is perceived and whether it can be trusted to operate alongside people.

Sensing contributes by defining the boundaries within which the system can act confidently. Perception is never perfect, and uncertainty must be explicitly accounted for in decision-making. Systems that treat perception outputs as fully reliable tend to behave aggressively under uncertainty, while systems that encode confidence and adjust behavior accordingly maintain safer and more predictable interactions.

These considerations lead to architectural choices that favor bounded behavior over maximum performance. Instead of optimizing for speed or expressiveness, systems are designed to operate within limits that ensure stability under variation. Movements may be slower, transitions more deliberate, and recovery behaviors more conservative, not because capability is lacking, but because safety requires maintaining control under imperfect conditions.

Human presence adds another dimension to this constraint, because safety is not only measured by physical outcomes but by perception. People evaluate safety through behavior that is legible and understandable. A system that moves abruptly, even if technically safe, can create discomfort. A system that hesitates excessively can appear uncertain. Both cases affect trust, which in turn affects whether the system is accepted in the environment.

This makes predictability as important as correctness. Actions must not only be safe but must appear safe as they unfold. The system's behavior should communicate intent through motion, timing, and consistency, allowing people to anticipate what will happen next without needing to interpret internal state.

Interruptibility becomes a critical property in this context. Humans must be able to stop or redirect the system easily, and the system must be designed to handle interruption without destabilizing. This requirement propagates through the architecture, affecting control strategies, task planning, and how state is maintained across actions.

Early pilots make these constraints visible. Systems that perform well in controlled demonstrations are often adjusted significantly when introduced into environments where people are present. Motion is slowed, force limits are reduced, and behaviors are simplified to ensure that interactions remain within acceptable bounds. These adjustments are not temporary compromises but reflections of the conditions under which the system must ultimately operate.

Over time, safety becomes a unifying constraint across all subsystems. It influences how sensing uncertainty is handled, how actuation is limited, how control is scheduled, and how tasks are structured. Treating safety as a separate layer leads to conflicts between performance and reliability, while embedding it into the architecture aligns these objectives from the beginning.

This integration also shapes how systems are evaluated. A system that performs well but behaves unpredictably will struggle to gain acceptance, while a system that operates consistently within known bounds can be trusted, even if it is less capable in absolute terms. Trust emerges from the alignment between behavior and expectation, not from peak performance.

As humanoids move from controlled environments into shared spaces, safety becomes inseparable from functionality. It defines not only what the system can do, but how it must do it, and under what conditions it can be allowed to operate. This shift sets the foundation for how systems are introduced into real workflows, where capability alone is not sufficient and reliability over time becomes the primary measure of value.

Reliability as an Architectural Outcome

As safety becomes embedded in how humanoids are designed and evaluated, attention shifts from what systems can do to how consistently they perform over time. Early progress is often framed around capability, whether a robot can walk, grasp, or execute a sequence of actions. Those milestones demonstrate feasibility, but they do not determine whether a system can be integrated into real environments. Reliability emerges from how those capabilities hold under repetition, variation, and duration, and that behavior is shaped by architectural decisions rather than isolated improvements.

Controlled demonstrations make capability visible because they minimize variability. Lighting is stable, objects are known, and interactions follow expected patterns. Systems are often reset between runs, allowing performance to be evaluated in short intervals where accumulated effects do not surface. Under those conditions, behavior appears consistent and repeatable, creating the impression that the system is ready for broader use. That impression changes once the same system is asked to operate continuously, without reset, and under conditions that cannot be fully controlled.

The transition from demonstration to pilot introduces a different set of requirements. Systems must perform the same task multiple times with small variations in setup, timing, and environment. They must carry forward the effects of prior operation, including thermal changes, calibration drift, and mechanical wear. The environment itself becomes less predictable, with objects shifting position, surfaces changing properties, and human interaction introducing interruptions that cannot be scripted in advance. These factors do not appear as isolated disruptions but as ongoing pressure on the system, revealing whether behavior stabilizes or degrades over time.

Performance that appears stable in a single run often begins to diverge when repeated. Movements that were precise at the beginning of a session may become less consistent as time passes. Timing margins narrow as compute and control loops absorb additional load. Grasp attempts require adjustment as perception and alignment drift slightly from their initial calibration. The system continues to function, but with increasing variability that affects how it is perceived and how it can be used.

Evaluation in these conditions shifts from whether a task can be completed to whether it can be completed consistently without introducing friction into the workflow. Buyers and operators are less concerned with peak performance than with predictability. A system that succeeds intermittently but requires supervision, resets, or manual correction creates operational overhead that limits its usefulness. A system that performs a narrower set of tasks consistently, even if more slowly, reduces that overhead and is easier to integrate into existing processes.

Consistency depends on how the system manages accumulated effects. Duration exposes interactions between sensing, actuation, and compute that are not visible in short runs. Thermal conditions change how components behave, affecting both responsiveness and precision. Sensors drift and require recalibration, sometimes subtly enough that errors appear only in edge cases. Mechanical wear alters dynamics, shifting how force is applied and absorbed. These changes do not occur independently, and their interaction determines whether the system remains within acceptable bounds.

Architectural decisions define how these effects are handled. Systems designed with explicit margins for timing, power, and thermal load are better able to maintain stable behavior under extended operation. Scheduling strategies that prioritize critical control loops over less time-sensitive processes prevent degradation from propagating across the system. Calibration procedures that can be applied without specialized intervention allow the system to recover alignment before errors accumulate. These choices do not eliminate variability, but they limit how that variability affects behavior.

Recovery behavior becomes central to reliability because deviation is unavoidable. A grasp may fail due to slight misalignment, or a movement may require adjustment due to unexpected contact. Systems that detect these deviations early, select bounded recovery actions, and return to a stable state without escalating the situation maintain continuity in operation. Systems that respond unpredictably, or that require external intervention to resolve minor issues, introduce disruption that compounds over time.

The design of recovery is tied closely to how failure is represented within the system. When uncertainty is encoded explicitly, through confidence measures or state estimation that reflects variability, behavior can be adjusted before errors propagate. When uncertainty is treated as noise to be ignored, small discrepancies can grow into larger failures that are more difficult to contain. Reliability emerges from acknowledging imperfection and designing mechanisms to manage it continuously.

Integration into workflows adds another layer of constraint. Systems must operate within processes that were not designed around them, often alongside human workers who expect predictable timing and behavior. Variability that might be acceptable in isolation becomes problematic when it affects downstream tasks or requires coordination across multiple steps. Reliability in this context includes not only the system's internal stability but also its ability to align with external processes.

As systems move toward deployment, expectations extend beyond technical performance to include uptime, maintainability, and predictability over extended periods. A system that requires frequent resets, manual calibration, or close supervision increases operational cost, even if its underlying capabilities are strong. A system that maintains stable behavior with minimal intervention reduces that cost and becomes more viable as part of a larger operation.

Economic considerations reinforce this shift. Inconsistent behavior introduces hidden costs through supervision, downtime, and reduced throughput, while consistent behavior allows for planning and integration. Architectural decisions that improve reliability often involve tradeoffs, such as limiting speed to maintain control margins or reducing model complexity to preserve timing guarantees. These tradeoffs affect both performance and cost, linking reliability directly to economic viability.

The relationship between capability and reliability explains why progress can appear uneven. Advances in perception, manipulation, or control may enable new behaviors, but those behaviors may not translate into operational value if they introduce additional variability or depend on conditions that cannot be sustained. Systems that expand capability without addressing consistency encounter friction when transitioning from demonstration to pilot, while systems that prioritize stability under variation are able to progress more steadily toward deployment.

Reliability, as a result, is not an outcome that can be added after capability is achieved. It is an architectural property that must be designed into the system from the beginning, influencing how subsystems interact, how resources are allocated, and how behavior is constrained under uncertainty. The systems that move toward real-world use are those that maintain consistent behavior across repetition, absorb variability without escalation, and operate within predictable bounds that can be trusted by the people and processes around them.

This perspective reframes what maturity looks like in humanoids. A system that performs impressively in isolated demonstrations but degrades under repetition remains limited in its practical value. A system that operates consistently, recovers from deviation without disruption, and maintains alignment with its environment begins to resemble a tool that can be relied upon. That distinction shapes how systems are evaluated, how they are funded, and how they are introduced into real workflows, linking architectural decisions directly to adoption.

Specificity in Humanoid Design

As reliability and safety begin to define what is acceptable in real environments, the discussion around capability shifts toward how broadly that capability can be applied. The idea of a general-purpose humanoid, able to perform a wide range of tasks across different environments, is often presented as the natural end state of progress. At the same time, many of the systems that move most quickly toward real-world use are designed around a narrower set of tasks, optimized for consistency within a defined scope. These two directions, generality and specificity, are not opposing philosophies but distinct architectural paths, each with its own implications for design, performance, and adoption.

General-purpose humanoids aim to operate across a wide range of tasks without requiring significant reconfiguration. This approach prioritizes flexibility, allowing a single system to adapt to different environments and workflows. Achieving this flexibility requires the system to handle a broad range of variation, from object properties and environmental conditions to interaction patterns with people. As a result, generality is not confined to intelligence or high-level reasoning. It extends across sensing, actuation, control, and system integration, with each layer required to support a wider operating envelope.

Expanding this envelope introduces complexity. Perception systems must interpret a wider variety of inputs under changing conditions. Manipulation must handle different shapes, materials, and force requirements. Control systems must remain stable across a broader range of timing and coordination demands. These requirements increase the number of edge cases the system must manage, and those edge cases tend to appear at the boundaries where reliability is most sensitive.

Task-specific humanoids follow a different path. They are designed to perform a defined set of tasks within a constrained environment, allowing the system to be optimized for consistency rather than flexibility. By limiting the range of variation the system must handle, architectural decisions can be tuned more tightly. Sensing can be calibrated to known conditions, manipulation strategies can be aligned with specific object types, and control can be optimized for predictable timing and motion patterns.

This constraint simplifies integration. With fewer variables to manage, systems can achieve stable behavior more quickly and maintain it more consistently over time. Reliability becomes easier to demonstrate, and safety margins can be defined more clearly. In early deployments, this often leads to faster adoption, because the system aligns more directly with existing workflows and requires less adaptation from the environment.

The tradeoff between these approaches becomes apparent when systems are evaluated under real conditions. General-purpose systems offer broader potential, but that potential must be realized across a wide range of scenarios, each introducing its own variability. Task-specific systems offer narrower capability, but that capability can be delivered with greater consistency, making it easier to integrate into operational settings.

Architectural decisions determine how this tradeoff is managed. Increasing generality typically requires additional sensing modalities, more complex models, and greater computational resources. These additions expand the system's ability to interpret and respond to variation, but they also increase power consumption, thermal load, and coordination complexity. Managing these factors requires careful integration, because improvements in one area can introduce constraints in another.

Task-specific architectures can allocate resources more selectively. By focusing on a defined set of tasks, systems can reduce the need for redundant sensing, simplify control strategies, and maintain tighter bounds on performance. This does not eliminate complexity, but it concentrates it in areas that directly support the intended use case, making behavior easier to predict and maintain.

The economic implications of these choices are significant. General-purpose systems tend to carry higher upfront complexity, both in hardware and software, as well as ongoing costs related to calibration, maintenance, and operation. Supporting a wide range of tasks requires continuous adaptation, which can increase the need for oversight and adjustment. These costs are justified if the system can deliver sufficient flexibility across multiple use cases, but they can be difficult to sustain if reliability varies across tasks.

Task-specific systems present a different cost profile. By limiting scope, they can achieve stable performance with lower operational overhead, reducing the need for frequent intervention or recalibration. This makes them easier to deploy within defined workflows, where consistency and predictability are valued. However, their narrower focus means that additional systems or reconfiguration may be required to handle different tasks, which introduces its own form of cost.

These dynamics influence how systems are adopted. In environments where tasks are well-defined and repetition is high, task-specific humanoids are often more practical, because they can deliver consistent performance within known bounds. In environments that require flexibility, general-purpose systems offer greater long-term value, but only if they can maintain reliability across variation.

The relationship between generality and reliability is not fixed. As systems improve, the boundary between these approaches can shift. Advances in sensing, control, and integration can allow general-purpose systems to handle variation more effectively, while modular architectures can enable task-specific systems to expand their capabilities without sacrificing stability. Progress in one area often informs the other, creating a gradual convergence rather than a strict division.

What remains consistent is that generality cannot be achieved independently of the rest of the system. Expanding capability requires corresponding improvements in integration, resource management, and operational design. Systems that extend generality without addressing these factors tend to encounter diminishing returns, where added flexibility introduces more variability than value.

In practice, early adoption is likely to favor systems that balance constrained generality with high reliability, expanding capability incrementally as integration improves. This approach allows systems to deliver value within defined contexts while building the foundation for broader use over time. As reliability increases and costs become more predictable, the range of tasks that can be supported expands, allowing general-purpose systems to emerge gradually from systems that were initially more specialized.

This progression reflects the underlying structure of humanoids as integrated systems. Decisions about how much generality to pursue and where to constrain capability are not only technical choices but architectural ones that shape performance, cost, and adoption. The systems that succeed are those that align these factors, delivering capability that is not only broad in potential but stable enough to be used in practice, and predictable enough to be trusted within real environments.

Generalization Is the Real Constraint

The natural question after reliability is whether a humanoid can perform the required tasks across different environments and conditions. Early demonstrations often suggest the answer is yes. A robot can grasp objects, navigate a room, or complete a sequence of actions under controlled conditions.

What becomes clear in practice is that capability alone is not the limiting factor. The real constraint is consistency. A system that performs well in one environment must maintain that performance across many environments, each with its own variability, unpredictability, and subtle differences. The challenge is not getting the robot to work once. It is getting it to work reliably across situations that cannot be fully anticipated.

A successful demonstration shows what is possible. Consistent performance shows what is usable. The gap between those two is where most of the difficulty resides.

The Variability of Real Environments

Real environments are not uniform. They differ in layout, lighting, materials, noise, human behavior, and countless other factors that are difficult to model exhaustively. Even small variations can have a significant impact on perception and control. A surface that reflects light differently can affect vision systems. A slightly shifted object can change the required grasp. A human moving unpredictably introduces new dynamics that must be handled in real time.

These variations are not edge cases. They are the norm. Any system intended to operate in human environments must account for this variability as a baseline condition, not as an exception. Designing for a fixed setting is relatively straightforward. Designing for a range of conditions that cannot be fully enumerated is a different problem.

This is where many systems begin to break down. What works in a controlled setting does not always transfer cleanly to a different environment. The assumptions embedded in the system no longer hold, and performance degrades in ways that are difficult to predict in advance.

Why Generalization Is Difficult

Generalization requires a system to interpret new situations using prior experience while maintaining stable behavior. This involves more than recognizing patterns. It requires the ability to adapt decisions in real time without introducing instability or unsafe behavior.

From a technical perspective, this places demands on every layer of the system. Perception must remain robust under changing conditions. Control systems must handle variation without oscillation or failure. Planning must account for uncertainty while still producing actionable outputs. These components are interdependent. Weakness in one layer can propagate through the system.

The difficulty is compounded by the fact that humanoids operate in environments designed for people, not machines. Objects are not placed for optimal sensing. Spaces are not arranged for predictable navigation. Interactions are not scripted. The system must operate within these constraints while maintaining performance.

This is fundamentally different from systems designed for controlled settings. In those cases, the environment can be structured to match the capabilities of the machine. With humanoids, the machine must adapt to the environment as it exists.

The Limits of Optimization

One approach to improving performance is to optimize the system for a specific set of tasks or conditions. This can produce strong results in narrow domains. However, optimization often comes at the expense of flexibility. A system that is highly tuned for one scenario may struggle when conditions change.

This creates a tension between optimization and generalization. Pushing too far toward optimization can reduce the system's ability to adapt. Pushing too far toward generalization can reduce efficiency and increase complexity. Finding the right balance is an ongoing challenge.

In practice, this means that progress is not linear. Improvements in one area can introduce new limitations in another. A more capable perception system may require additional compute, which affects power and thermal constraints. A more flexible control system may introduce variability that needs to be managed carefully.

The result is a system that must be designed with tradeoffs in mind, where gains in one dimension are evaluated against their impact on the overall behavior.

From Demonstrations to Deployment

The difference between demonstration and deployment becomes especially clear when systems are placed in real environments. A demonstration is typically structured to highlight success. Conditions are selected, variables are controlled, and the sequence of actions is known in advance.

Deployment removes these constraints. The system must operate continuously, handle unexpected situations, and recover from errors without external intervention. It must maintain performance over time, not just during a short sequence.

This is where generalization becomes critical. A system that cannot handle variation will require constant supervision or frequent adjustments. This limits its usefulness and increases operational cost. For humanoids to be practical, they must reduce the need for intervention, not increase it.

The transition from demonstration to deployment is therefore not just a scaling problem. It is a change in the nature of the problem itself. What worked in a controlled setting must now work under conditions that are only partially known.

Stability, Predictability, and Trust

Consistency is closely tied to trust. For a system to be accepted in human environments, it must behave in a way that is stable and predictable. Users need to understand how the system will respond under different conditions. Unexpected behavior, even if technically correct, can reduce confidence.

This places additional requirements on generalization. It is not enough for the system to adapt. It must adapt in a way that preserves expected behavior. Changes in performance should be gradual and understandable, not abrupt or opaque.

This is particularly important in environments where humans and machines interact directly. Small inconsistencies can have outsized effects on perception and acceptance. A system that behaves reliably builds trust over time. A system that behaves unpredictably, even occasionally, can undermine that trust quickly.

Designing for predictability while maintaining adaptability is one of the central challenges in humanoid systems.

Generalization as the Bottleneck to Scale

As the field moves toward broader deployment, generalization becomes the primary constraint on scale. Hardware can be replicated. Components can be manufactured at volume. Capital can support expansion. But if systems cannot operate consistently across environments, scaling remains limited.

This constraint is not always visible in early stages. It becomes apparent as systems move out of controlled settings and into more diverse environments. Each new setting introduces variables that must be handled without degrading performance.

Addressing this requires a combination of approaches. Data must capture a wide range of conditions. Models must be trained to handle variation. Systems must be designed to fail safely and recover effectively. None of these are isolated solutions. They must work together.

The path forward is therefore not defined by a single breakthrough, but by incremental improvements across multiple layers that collectively increase robustness.

Understanding generalization as the core constraint changes how progress is measured. Success is not defined by isolated achievements, but by the ability to maintain performance across conditions. A system that performs slightly less efficiently but consistently across environments may be more valuable than one that achieves higher performance in a narrow setting.

This perspective also changes how milestones are interpreted. Early demonstrations remain important, but they are no longer sufficient indicators of readiness. The focus shifts to repeatability, reliability, and adaptability.

Over time, this leads to a more grounded understanding of the field. Progress is still real, but it is evaluated through the lens of what can be sustained in practice, not just what can be achieved under ideal conditions.

5

Funding Human-Like Machines at Scale

For a long time, it was tempting to believe that humanoids would follow the same capital path as software and AI models. Progress felt familiar. Demos improved. Capabilities expanded. Valuations moved ahead of deployments. From a distance, the arc looked recognizable enough that many assumed the funding logic would sort itself out.

What I saw instead was a growing mismatch between how capital expected progress to unfold and how physical systems actually mature.

That mismatch was structural. The moment humanoids moved beyond isolated demonstrations and into prolonged interaction with the physical world, the assumptions embedded in traditional venture funding began to strain. Timelines stretched. Costs behaved differently. Risk surfaced in places spreadsheets were not designed to model. Progress came in uneven steps rather than smooth curves.

This was not a problem of ambition or effort. Founders, builders, and researchers were doing what the work required. The problem was that the financial frameworks surrounding them had been shaped for assets that scale through abstraction, not systems that accumulate wear, require calibration, and reveal their limits only through contact with reality.

I saw this firsthand. Capable companies struggled not because they lacked technical insight, but because they were forced to operate inside expectations that no longer matched their fields. Capital wanted compounding returns before the machines could sustain compounding reliability. Markets looked for signals that humanoids were not yet structurally able to produce. Valuation logic borrowed from software distorted incentives at exactly the wrong phase.

Over time, what became clear to me was that humanoids did not just need better technology. They needed a different relationship with capital, one that understood physical risk, integration debt, and the slow conversion of capability into usefulness. Funding was not simply fuel for growth. It was part of the architecture.

Only once funding begins to respect physical reality will human-like machines stand a chance of scaling beyond promise.

When Capital Meets Physical AI

Capital behaves differently when it meets physical systems. In software, money mostly accelerates iteration. In humanoids, money changes the shape of the problem itself. It determines which risks can be retired, which must be carried longer, and which quietly compound if ignored.

In software-first AI, capital primarily funds people and compute. The assets are intangible, the iteration loops are short, and mistakes can often be rolled back with limited consequence. In physical AI, capital funds time, materials, tooling, suppliers, test infrastructure, and service readiness. Many of these costs are front-loaded, irreversible, and coupled to long lead times that do not respond to urgency. Once metal is cut or a supplier is locked, capital buys commitment.

It is tempting to talk about humanoids as an extension of AI progress, where better models and more data eventually smooth everything out. That framing collapses under physical reality. Models improve behavior, but they do not shorten supply chains, validate safety cases, harden components, or build field operations. Those outcomes require capital to be deployed with intent, sequence, and restraint.

The First Collision with Physical Reality

The first collision between capital and physical AI usually happens earlier than founders expect. A team may raise enough money to build an impressive prototype, only to discover that the next step is not incremental. Moving from a lab system to something that can operate reliably around people requires a different spending approach entirely. Safety systems, redundancy, test coverage, and documentation suddenly dominate the budget. These are prerequisites for entry into real environments.

This shift exposes a common misunderstanding: Many people assume that more capital always reduces risk. In humanoids, poorly timed capital can increase it. Large checks deployed before a system's failure modes are understood tend to mask fragility rather than resolve it. Founders hire ahead of clarity, lock designs prematurely, or expand scope before discipline sets in. The burn rate rises, but the underlying uncertainty does not fall at the same pace.

Physical AI forces capital to retire risk, not pursue ambition. Each phase has a dominant class of uncertainty, and money is only productive when it is aimed squarely at removing that class. Early capital must fund integration, not scale. Mid-stage capital must fund repeatability, not expansion. Later capital must fund manufacturing stability and service infrastructure, not novelty. When capital skips ahead of the system's maturity, the company accumulates hidden debt in the form of rework, field failures, and eroded trust.

How Physical Risk Compounds

Physical risk compounds differently than software risk. In software, a bug may affect many users, but it rarely damages the underlying platform permanently. In humanoids, repeated small failures degrade components, strain service teams, and poison customer confidence. A grasp that slips occasionally in a demo becomes unacceptable in a real environment. A balance recovery that looks fine on video becomes a liability when repeated over hours of operation.

This changes how capital efficiency should be measured. Traditional metrics like headcount growth or feature velocity say very little about progress in physical AI. The meaningful indicators live elsewhere. Yield stability, fault class reduction, mean time to repair, service readiness, and safety margin consistency are the signals that capital is doing its job. These metrics tend to improve slowly at first, then compound once the system stabilizes. Capital that chases optics instead of these fundamentals rarely produces durable value.

There is also a psychological dimension to this. Founders and investors trained in software are used to interpreting momentum as acceleration. In humanoids, momentum often looks like restraint. It shows up as fewer configuration changes, narrower task focus, and longer validation cycles. Capital must support this discipline rather than punish it. When funding structures reward rapid expansion before the system is ready, they actively work against the physics of the problem.

Money shapes behavior. In physical AI, the wrong incentives can push founders to outrun their own understanding. The right incentives encourage founders to slow down precisely when patience creates leverage.

Over time, companies that internalize this dynamic begin to treat capital as an instrument rather than a buffer. Each raise is tied to specific retired risks, not to abstract milestones or market narratives. Spending plans are built around test coverage, supplier validation, and service learning curves rather than feature road maps alone. Progress is communicated in terms of reduced uncertainty, not increased visibility.

From the outside, this kind of company can look conservative. From the inside, it is intensely focused. The founders understand that physical AI does not reward bravado. It rewards alignment between money, physics, and operations. Capital that respects this alignment becomes a force multiplier. Capital that ignores it becomes a source of instability.

When capital meets physical AI with discipline rather than impatience, progress becomes slow, then steady, then difficult to stop.

Why Physical AI Requires a Different Risk Model

For a long time, the dominant intuition around risk in AI came from software. Models improved with data. Failures informed retraining.

Scale reduced marginal cost and increased reliability. Even when progress was uneven, the direction of travel felt predictable enough that risk appeared to compress over time.

What shifted my view was not a single failure or setback, but the accumulation of small, stubborn realities that only appeared once intelligence was embodied. As soon as systems began interacting with the physical world, risk stopped behaving like something that could be averaged away. It became something that surfaced irregularly, accumulated locally, and demanded attention in ways that software-trained capital was not accustomed to pricing.

The key difference in these physical fails lies in their operation. Physical AI introduces forms of risk that do not smooth out with scale. Instead of diminishing, some risks become more visible as exposure increases. Wear accumulates. Edge cases multiply. Interactions diversify. The system does not simply get better by seeing more data. It reveals more of what it cannot yet handle. That revelation is a property of operating in environments that refuse to converge.

In physical systems, iteration carries a cost every time. Components age. Calibration shifts. Recovery consumes time and attention. Learning still happens, but it is constrained by economics, safety, and liability. Each iteration leaves a mark.

Early gains in capability do not reduce risk uniformly. They often expose new classes of risk that were previously dormant. A humanoid that can now grasp more objects encounters more surface variability. A system that moves faster encounters tighter stability margins. A machine that operates longer encounters thermal and wear-related effects that short runs never reveal. Capability expands, but so does the surface area where things can go wrong.

When Capital Misreads Progress

Capital trained on software curves often expects risk to compress as competence increases. In physical AI, competence frequently precedes risk visibility. What looks like regression is often exposure. What looks like instability is often honesty. Systems that are still sheltered appear safer than systems that are actually learning through contact with reality.

This mismatch has consequences. Founders are pressured to demonstrate smooth progress when the underlying system is revealing uneven truths. Builders are incentivized to hide variance rather than instrument it. Researchers are asked to explain away behaviors that are simply the cost of operating in the real world. None of this improves outcomes. It only delays alignment.

The cost of a single failure can outweigh the benefit of many successes. This asymmetry reshapes acceptable risk thresholds and slows adoption until confidence is earned deliberately. This concentration of consequences constrains liabilities.

Liability reshapes technical decisions. In software, failure is often reputational. In physical systems, failure can be contractual, regulatory, or insurable. That reality feeds back into design choices, operating envelopes, and deployment pacing. Risk is no longer abstract. It is priced, documented, and negotiated. Progress that ignores this layer remains theoretical.

How Markets Recalibrate

Behavior changes once founders and investors take reality into account.

Learning still compounds in physical AI, but it compounds through disciplined exposure, careful scoping, and shared understanding of failure modes. It compounds when systems are instrumented to surface problems early rather than mask them. It compounds when organizations accept that risk reduction is not linear and that stability emerges through integration rather than optimization.

Over time, markets begin to recalibrate. Early capital often prices physical AI as if it were delayed software. Later, capital prices it as an evolving system with known classes of uncertainty. The difference between those two perspectives determines which companies survive the middle phase of development. Those that align capital with the true risk profile gain room to mature. Those that do not are forced into narratives that reality eventually contradicts.

I recognized that this risk profile was not a temporary anomaly. It was the defining feature of the category. Physical AI promises grounded progress that must be earned through exposure, coordination, and restraint.

Understanding this reframes how success should be judged. The question is not whether risk exists, but whether it is being surfaced, bounded, and reduced in ways that organizations can plan around. Systems that make risk legible progress faster in the long run than systems that appear smooth but conceal fragility.

This is why physical AI forces capital to learn again. It demands a shift from expecting acceleration to valuing absorption, from celebrating abstraction to respecting contact, and from assuming that intelligence alone will smooth the curve. Risk does not disappear in humanoids. It changes shape. Only once that shape is understood can progress be funded without distortion.

The Economic Interface Between Humans and Humanoids

The economic interface is the mechanism through which physical work becomes legible, contractible, and accountable inside human organizations. For humanoids, this interface is not optional once capital meets physical reality. It is the difference between a machine that impresses and a system that earns its place in an operation.

The reason this matters is simple. Human environments already have an economic structure. Work is priced, scheduled, supervised, audited, insured, and improved over time. People fit into that structure because their capabilities, limits, and costs are well understood. When humanoids enter those environments, they do not arrive as neutral tools. They arrive as new actors that must integrate into existing flows of responsibility and value. The economic interface is where that integration either happens cleanly or breaks down.

Early conversations around humanoids tended to focus on form and capability. Could the robot walk? Could it grasp? Could it reason? Those questions were necessary, but incomplete. Capability mattered only when it could be translated into predictable outcomes that others could plan around.

An economic interface turns physical action into something a manager can schedule, a finance team can budget, a safety officer can approve, and an operator can trust. Without it, even impressive machines remain trapped in pilot mode, because no one downstream can rely on them without constant human mediation.

What the Economic Interface Actually Does

In practice, this begins with how work is defined. Humans think in tasks that map to intent and outcome. Move this. Place that. Hold steady while something else happens. An economic interface must present work at that level while still carrying the constraints that make it safe and feasible. It needs to encode what success looks like, what failure looks like, and what happens when conditions fall outside the expected range.

I saw repeatedly that systems requiring constant translation by a specialist never scaled. When a robot needed an engineer to interpret logs, adjust parameters, or explain behavior after every exception, it created a bottleneck that no amount of enthusiasm could overcome. By contrast, when a system could explain itself in operational terms, adoption followed naturally.

That requirement reshaped how I thought about autonomy. What matters is not independence, but reliability within agreed boundaries. An economic interface defines those boundaries explicitly. It makes clear which actions the system can take on its own, which conditions require escalation, and how that escalation occurs without disrupting the surrounding workflow.

This is where humanoids diverge sharply from other robots. Fixed automation is typically embedded into a tightly controlled process. Its interface is designed once and rarely revisited. Humanoids, by contrast, operate in spaces designed for people, where variation is constant, and adaptation is expected. The interface must therefore support negotiation rather than enforcement. It must allow humans to intervene, override, or reassign work without breaking the system.

I observed that the most fragile deployments were those that treated humanoids as drop-in labor. They promised substitution without redesigning the interface through which work was requested and validated. The result was confusion on both sides. Humans expected flexibility that the system could not safely provide, and the system encountered edge cases it was never meant to handle.

The stronger deployments took a different approach. They treated the humanoid as a new category of worker with defined capabilities and limits. Tasks were scoped deliberately. Interfaces were built to reflect real constraints rather than aspirational ones. Over time, as reliability improved, those scopes expanded. This progression was not driven by breakthroughs in intelligence alone, but by interfaces that allowed the capability to be translated into value incrementally.

How Work Gets Defined and Measured

Measurement sits at the core of this translation. Physical work only becomes economic when it can be measured in ways that matter to the organization. That includes time, energy, wear, error rates, and recovery cost. An economic interface must capture these signals automatically and present them in forms that different stakeholders can use. Operations teams need status and exceptions. Finance teams need cost per task and variance. Safety teams need evidence that limits were respected.

This evidence cannot be an afterthought. If measurement is bolted on later, it rarely aligns with how decisions are actually made. When measurement is designed into the interface from the start, it shapes behavior upstream, influencing how tasks are defined, how failures are categorized, and how recovery paths are exercised rather than improvised.

In early humanoid programs, risk often felt vague. Something might go wrong, but it was unclear what, when, or how costly it would be. An economic interface breaks that uncertainty into parts. It shows which risks are frequent but cheap, which are rare but expensive, and which are unacceptable under any circumstances. That clarity allows organizations to make informed tradeoffs rather than blanket rejections.

The interface also governs how humans and humanoids coordinate in time. Human work is scheduled around shifts, deadlines, and dependencies. A system that cannot express its availability, duration, or uncertainty creates cascading inefficiencies. For example, a robot that could not estimate how long a task would take forced supervisors to buffer excessively, reducing overall throughput. When estimates improved, planning improved alongside them.

An economic interface that exposes retries as economic events, rather than silent internal loops, changes how tasks are structured and prioritized. It encourages designs that reduce variance rather than chase peak performance.

Accountability and Trust

In human organizations, every attribution supports training, compliance, and improvement. Humanoids complicate this because agency is distributed across software, hardware, and operators. An economic interface must reconstitute accountability in a way that fits existing norms. It must be clear who requested a task, which system executed it, under what conditions, and with what outcome.

Without this clarity, audits become painful, and trust erodes. With it, humanoids become legible participants in regulated environments. Deployments accelerate once compliance teams answer questions without improvisation. The robot did not need to be perfect. It needed to be inspectable.

The economic interface is also where human trust is built or lost. People are remarkably tolerant of limitations when they are clear and consistent. They are far less tolerant of unpredictability. An interface that sets expectations accurately, even if conservatively, earns trust. One that overpromises and underdelivers burns it quickly. Trust, once lost, is expensive to regain.

This has implications for how companies communicate externally as well. Marketing that outruns the interface creates pressure on deployments that the system cannot sustain. Communication that reflects the interface reality aligns sales, operations, and engineering.

How the Interface Shapes Progress

Pricing models depend on the interface. Physical work cannot be sensibly priced without knowing what is being delivered and how consistently. Per-task pricing, per-hour pricing, and subscription models each require different proofs. An economic interface provides those proofs. It shows not just that work was attempted, but that it met defined criteria under defined conditions.

This pricing clarity feeds back into design decisions. If a task consistently consumes more energy or wears out faster than expected, pricing can reflect that reality. Customers can then choose whether the task is worth automating or whether upstream changes would improve economics. Without this feedback loop, costs remain hidden until they surface as margin erosion or service failures.

One of the more subtle effects I noticed was how interfaces shape organizational behavior. When a humanoid system exposes clear metrics and recovery paths, humans adjust their workflows to complement it. They place objects more consistently and schedule tasks more deliberately. The system improves because the interface encourages better coordination.

As organizations deploy multiple systems, consistency becomes critical. An interface that behaves differently at each location multiplies training cost and error. An interface that behaves consistently allows scale without linear increases in overhead.

The economic interface also defines upgrade paths. Physical systems improve over time, but upgrades carry risk. An interface that supports versioning, rollback, and compatibility allows improvements to be introduced without disrupting operations. An interface that lacks these mechanisms forces organizations to choose between stagnation and instability. In my experience, most choose stagnation, and progress stalls.

Throughout this process, my own perspective shifted. I stopped evaluating humanoid efforts based on demos or road maps and started evaluating them based on interface maturity. Could the system explain itself? Could it be priced? Could it be audited? Could it recover gracefully? These questions turned out to be better predictors of success than any single technical benchmark.

At a higher level, the economic interface reframes what progress looks like. Progress is not just more capability. There is better alignment between capability and value. It has a narrower variance. It is clearer contracts. There are fewer surprises. These changes may not make headlines, but they change purchasing behavior.

As capital began to understand this, funding patterns shifted. Investors who grasped the importance of the interface asked different questions. They wanted to see task definitions, service metrics, and recovery costs. They cared less about speculative breadth and more about repeatable depth. That shift did not happen overnight. It happened through exposure to systems that either failed or succeeded at this translation.

Humanoids will succeed because they fit into human systems cleanly enough to earn trust, deliver value, and allow capital to be allocated, priced, and scaled without distorting the work itself. The economic interface is where that fit is negotiated, enforced, and improved over time.

The Phase Where Most Humanoid Companies Stall

Once the economics define progress, a pattern starts to repeat across humanoid efforts that initially look strong. The failures that follow are rarely sudden, and they are almost never the result of a single technical shortcoming. They emerge during a specific phase, after early demonstrations succeed but before real integration stabilizes, when expectations rise faster than the system can absorb variation.

I began to recognize this phase because companies kept working while momentum quietly leaked away. Teams grew. Road maps expanded. Capital was raised. From the outside, everything looked intact. Internally, however, the system was accumulating stress faster than it was resolving uncertainty, and the interface between capability and value had not yet matured enough to carry that load.

This phase typically begins when a humanoid transitions from controlled demonstrations into broader exposure, even if that exposure is still limited. Tasks are extended slightly. Environments become less predictable. Stakeholders outside the core technical group begin to interact with the system. At that point, the absence of a robust economic interface starts to matter. Work can be performed, but it cannot yet be explained, priced, scheduled, or recovered from consistently.

Progress does not stop. In many cases, capability continues to improve. Models get better. Motion becomes smoother. Perception handles more edge cases. These gains are real, but they do not translate cleanly into operational confidence. Instead, each improvement introduces new dependencies, new failure modes, and new coordination costs that the surrounding organization is not prepared to manage. The pattern reveals what companies choose to measure and what they choose to ignore.

The Metrics That Mislead

I watched companies enter this phase while still believing they were on track. They measured success through demonstrations, customer interest, or follow-on funding, rather than through operational stability. Because failures were rarely catastrophic, they were treated as anomalies rather than signals. Recovery required human intervention, but that intervention was normalized instead of interrogated. Over time, the gap between what the system could do and what the organization claimed it could do widened.

Without a mature interface, costs become difficult to attribute. Downtime is absorbed quietly. Wear accumulates without being priced. Engineering effort shifts from advancing capability to firefighting exceptions. None of these issues appears clearly on a balance sheet, but they erode the conditions required for scale.

This phase rewards the wrong optimizations. Companies learn how to make demos more robust, how to reset systems faster, and how to hide uncertainty behind conservative behavior. These adaptations improve surface reliability but slow the development of the interface itself. Work continues, but learning stalls, because failures are patched rather than structured.

How Companies Escape the Stall

At the same time, external expectations harden. Customers begin planning around promised capabilities. Investors expect timelines to compress. Hiring expands to meet perceived demand. The system, however, is still operating with narrow margins and incomplete translation between action and value. When reality pushes back, it does so unevenly, creating friction across sales, engineering, operations, and support.

This phase ends with exhaustion. Founders burn time resolving the same classes of issues repeatedly. Capital becomes harder to deploy effectively because each dollar funds maintenance rather than progress. Confidence erodes, not because the technology failed, but because it never became legible enough to support the promises built on top of it.

What separates companies that survive from those that stall is not technical brilliance alone. It is the decision to slow down at this inflection point and invest in the interface rather than pushing past it. The companies that reset expectations, narrow scope, and formalize how work is defined, measured, and recovered from are able to convert capability into durable value. The ones that do not remain trapped in a perpetual pilot state: impressive but economically fragile.

How Capital Markets Learn

Capital markets do not misunderstand humanoids because markets learn from patterns, and humanoids break the patterns capital markets use to read. They produce different signals, often in the wrong order. Humanoids do not produce those signals early.

The Signals That Confused Early Capital

When humanoid companies begin to show real progress, the first visible change is rarely revenue or margin. It is constraint discovery. Systems reveal where they break. Interfaces get narrower. Deployment scopes shrink before they expand. From inside the work, this looks like maturity. From the outside, it can look like a retreat.

Companies that were learning the fastest often appeared to be slowing down. Companies that looked smooth were often deferring the reckoning. Capital, trained to reward momentum, initially favored the wrong signals.

Capital markets learn through exposure to repeated outcomes. They learn when similar companies hit similar limits at similar stages, and when the ones that survive do so by making comparable adjustments. Over time, pattern recognition replaces optimism.

In humanoids, that pattern took longer to form because progress did not move in straight lines. A company might demonstrate impressive capability one quarter, then deliberately constrain its system the next. Another might slow deployments while improving instrumentation, safety, or interface maturity.

Early capital often misread those decisions. Later, capital began to recognize them as signals of discipline.

How Markets Learned to Read Differently

One of the strongest indicators was a shift in the questions asked. Investors began asking where not to operate. Instead of pushing for aggressive timelines, they asked how exposure was being staged. They focused on task definition, recovery behavior, and deployment economics.

Markets also learned to distinguish between different kinds of progress. Demonstrations stopped being treated as milestones in their own right. What mattered was whether demonstrations led to tighter interfaces, clearer pricing, lower variance, and fewer surprises in the next phase. Progress was no longer about what a humanoid could do once. It was about what it could do repeatedly without explanation.

That distinction reshaped valuation logic. Early valuations often assumed breadth would arrive quickly and justify scale. Later valuations favored depth, especially in constrained domains where learning compounded. Companies that could show narrowing uncertainty, even with a limited scope, began to command more durable confidence than companies chasing generality without containment. Markets started to reward companies that entered the world carefully but deliberately, rather than those that stayed abstract for too long.

This also changed how companies viewed patience, supporting structured exposure, accepting uneven progress, and allowing systems to mature without forcing premature narratives. Capital that understood this stopped pushing for acceleration everywhere and started asking where acceleration was actually safe.

Over time, this learning reshaped who participated. Some investors stepped back when it became clear that humanoids would not behave like software. Others leaned in once they understood that the risk profile, while different, was manageable if approached correctly. The market did not converge on a single view, but it did segment more clearly.

Companies aligned with capital that understood physical learning cycles gained room to mature. Companies mismatched with their capital base often found themselves overextended, under-instrumented, or forced into decisions that compromised long-term viability.

Secondary signals began to matter more. Customer renewal behavior, deployment expansion within a single site, and reduced need for specialist intervention became indicators that markets trusted more than headline capability claims. These signals were quieter, but they were harder to fake.

As these patterns accumulated, the narrative around humanoids shifted. Progress was no longer framed as imminent replacement or sudden transformation. It was framed as gradual integration. That framing stabilized interest.

By the time markets began to understand humanoids on their own terms, much of the early noise had already passed. The companies still standing were not necessarily the ones that had promised the most, but the ones that had aligned capital, interface, and risk in a way that allowed learning to continue.

What determines outcomes is how much damage occurs before capital adapts. In humanoids, shared language around interfaces, risk, and deployment shortened that learning curve.

That is how capital markets learn about humanoids' progress. Not through hype cycles or singular breakthroughs, but through repeated exposure to systems that respect physical reality and make their learning visible.

6

Making Humanoids Culturally Acceptable

CULTURE IS WHERE humanoids will be accepted or refused. Not in labs, not in boardrooms, but in the spaces where people live and work.

For most of the public, humanoids have remained an abstraction, but never encountered directly. That distance allowed the conversation to stay theoretical. It no longer can. The technology has matured enough that presence is no longer hypothetical. People are beginning to ask what it would mean to share space with machines shaped like themselves.

The common concern is replacement. After decades of automation anxiety and recent waves of AI-driven disruption, that fear is understandable.

The short answer is no. The longer answer requires understanding what humanoids actually are, what they demand from organizations, and what remains irreducibly human even as capability grows. That understanding reframes the tension, moving from replacement to integration, from fear to negotiation, from abstraction to presence.

Why the Fear of Human Replacement Dominates

The fear that humanoids will replace humans dominates public conversation because it aligns with how people have learned to interpret technological change, economic uncertainty, and power asymmetry. The concern surfaces not because people have examined humanoid deployments closely, but because the idea fits a story that has been told repeatedly, reinforced culturally, and rarely corrected with better information.

Humanoids tend to collapse multiple anxieties into a single symbol. Economic displacement, loss of agency, distrust of institutions, and unease about automation all converge when a machine takes human form. Unlike earlier technologies that altered work indirectly, humanoids appear to completely replace humans. That visual proximity makes abstract concerns feel personal.

How the Story Builds

Media amplification plays a central role in this framing. Headlines gravitate toward extremes because extremes attract attention. Stories that suggest mass replacement, total automation, or imminent job loss spread faster than nuanced accounts of constrained deployment and gradual adoption. Over time, repetition hardens speculation into assumed trajectory. Each new demo or headline then feels like proof of what was already expected.

Another driver is the long shadow cast by earlier automation results. The twentieth century produced real instances of job loss tied to mechanization, and those memories are passed forward culturally even when the specifics no longer apply. People remember the pain without retaining the context. They recall factories shrinking, towns hollowing out, and livelihoods disappearing, but not the economic, geographic, and policy decisions that shaped those outcomes. Humanoids become the next vessel for unresolved history.

This inherited anxiety is reinforced by the way work itself has changed. In many economies, employment has become less stable, benefits have weakened, and career paths have flattened. When work already feels fragile, any new technology is interpreted as a threat rather than a tool. Humanoids arrive into that environment carrying the burden of mistrust built by decades of unrelated decisions.

The human form factor intensifies this reaction in ways other machines do not. A humanoid standing upright triggers social and psychological responses that evolved for interaction with other people. The reaction is instinctive rather than irrational.

This is compounded by a tendency to overestimate the pace of technological change while underestimating the friction of adoption. Most people unfamiliar with new technology have no reason to know how slowly these physical technologies actually spread. In reality, physical technologies propagate unevenly, constrained by capital cycles, regulation, training, and cultural acceptance.

I have watched impressive demonstrations be interpreted as proof of inevitability, even when those demonstrations were the result of careful staging, narrow task definition, and extensive human support. The audience never sees the scaffolding. Without visibility into that scaffolding, it is easy to assume that replacement is simply a matter of time.

Economics are also easy to misread. When people hear that machines can work continuously, do not require wages, and do not tire, they extrapolate linearly to job elimination. What is missing from that reasoning is the cost of making machines reliable, safe, and accountable in real environments. The comparison is against a machine that does not yet exist.

Power dynamics also shape perception. Many people do not fear humanoids themselves. They fear how decision-makers might use them. The concern is not about what robots can do, but about what employers might choose to do with them. Humanoids become a proxy for broader distrust in corporate and institutional motives.

The fear is further amplified by how success is measured and reported. Metrics like number of tasks completed, hours operated, or objects manipulated are presented without context about supervision, failure rates, or intervention frequency. Without that context, the numbers suggest machines operating alone. Over time, they reinforce the idea that humans are being edged out, even as human involvement remains central.

This distrust often begins with technology but quickly move to fairness, access, and control. People ask who benefits, who decides, and who bears the cost when things go wrong. Without credible answers, fear fills the gap.

There is no easy way to learn the inner workings of technological advancement. There is little shared literacy around how physical

AI systems are actually developed, deployed, and constrained. Most people encounter humanoids through short clips or speculative commentary rather than through sustained exposure to real use cases.

Even well-intentioned advocates sometimes worsen the problem by overpromising. Claims about general intelligence, full autonomy, or human equivalence may excite investors or audiences in the short term, but they also validate replacement fears. When the narrative runs ahead of reality, trust erodes on both sides. Skeptics feel misled, and supporters are forced to defend claims that cannot yet be delivered.

Why the Fear Persists

Change is hard to picture clearly. It is easier to picture jobs disappearing than to picture roles evolving in subtle ways. Displacement is a simple story that can be told quickly and remembered easily.

There is also a mismatch between individual and system-level perspectives. At a societal level, humanoids may absorb labor shortages or reduce injury. At an individual level, someone may still worry about their specific role. Without deliberate communication and transition planning, fear appears rational.

I have seen this disconnect firsthand. Organizations talk about productivity gains and safety improvements, while workers worry about schedules, retraining, and relevance. When those conversations are not aligned, technology becomes an antagonist rather than a support. The fear persists not because people reject progress, but because they do not see themselves in it.

Timing becomes the key factor in implementation. They are arriving during a period of geopolitical instability, economic polarization, and rapid AI advancement. Trust in institutions is low. When multiple sources of uncertainty converge, new technologies inherit suspicion by default. Humanoids arrive into a climate primed for worst-case interpretations.

This context explains why technical reassurances alone do not dissolve the fear. Explaining limitations, timelines, or safeguards addresses logic but not lived experience. Fear is sustained socially, not individually. It is reinforced through conversation, repetition, and shared assumptions. Countering it requires sustained, visible evidence over time, not arguments in isolation.

What ultimately sustains the replacement narrative is that it is rarely challenged by concrete alternatives. Without a credible picture of how human roles change, fear remains the dominant storyline. The absence of a positive, grounded narrative leaves a vacuum that replacement fills.

This is the real cultural challenge facing humanoids. It is not enough for the technology to be safe or useful. People must understand where they fit, how work evolves, how dignity is preserved, and how agency is maintained. Until that visibility exists, fear will continue to dominate regardless of technical progress.

The persistence of this fear does not mean it should be indulged, but it should be taken seriously. Ignoring it creates backlash. Addressing it requires acknowledging why it makes sense to people given the signals they receive. Only then can alternative frames take hold.

Humanoids Will Not Replace Humans

This fear is the most prevalent in humanoid deployment, and it is understandable. The belief that humanoids will replace humans is what most people intuit the moment they see a human-shaped machine walk, grasp, or speak. That intuition, however, collapses under examination.

From what I've seen, humanoids will change human work. But replacement, in the literal sense most people imagine, is structurally implausible due to the economic, organizational, and physical realities of humanoid deployment.

The first mistake is the assumption that human labor is a single interchangeable category. In reality, human work is fragmented across roles, contexts, and accountability structures that evolved over centuries. Jobs are arrangements of responsibility, judgment, liability, adaptability, and social coordination. Even highly repetitive roles include exception handling, informal decision-making, and context sensitivity that rarely appear in job descriptions but dominate day-to-day reality.

Humanoids must be specified, constrained, supervised, insured, and justified. They do not arrive as free agents. They arrive as capital assets embedded inside organizations that already carry legal obligations, labor agreements, safety regimes, and cultural norms. Every deployment decision is filtered through these layers before a single task is automated.

Projecting software-style scaling dynamics onto physical systems is also extremely difficult to achieve at large scale. In humanoids, every unit carries material cost, maintenance burden, and operational risk. Each additional deployment expands the surface area where things can fail, not just the upside where value can be captured.

I've seen how narrow the viable operating envelope actually is. Systems that look general in demos reveal tight constraints once placed in real environments. Tasks must be carefully scoped. Interactions must be choreographed. Recovery procedures must be planned in advance. Even then, human oversight remains integral, not as a temporary crutch, but as part of the system design. Replacement assumes independence. Real deployments demand integration.

What History Actually Shows

Historically, labor-displacing technologies that truly replaced humans removed entire categories of work rather than mimicking human form. The elevator eliminated elevator operators because it changed the interface to vertical transport. The ATM reduced teller headcount because it restructured access to cash. In both cases, the technology altered the system around the work.

Humanoids, by contrast, are designed precisely because the system around the work is not being redesigned. They are an adaptation to human-built environments, not a rewrite of them.

Technologies that replace humans tend to simplify systems. Humanoids are introduced precisely because systems remain complex, variable, and resistant to simplification. Their value proposition depends on that complexity continuing to exist.

Accountability is another completely human characteristic still needed in work. Human workers carry legal and moral responsibility in ways machines cannot. When something goes wrong, organizations rely on human judgment, escalation, and explanation to resolve ambiguity. Humanoids cannot absorb accountability. They can only operate within accountability structures designed and maintained by humans. In regulated environments—healthcare, logistics, manufacturing—compliance regimes assume human agency in the loop. Humanoids add supervision layers rather than eliminating them.

From a purely economic perspective, replacement narratives also ignore cost structure. Human labor is expensive in some contexts and cheap in others. Humanoids are capital-intensive everywhere. Their economics only work when they are deployed into tasks that are dull, dangerous, or physically taxing enough that human cost is already high or rising. Even then, the comparison is not one humanoid to one human. It is one humanoid plus support infrastructure versus a team of humans operating with flexibility.

What I observed repeatedly is that humanoids tend to absorb the least human parts of jobs rather than the jobs themselves. They take on the segments of work that humans actively try to avoid, where injury risk, fatigue, or monotony are highest. This shifts the human role rather than eliminating it. Humans move upstream into coordination, exception handling, and quality control. The job changes shape, but it does not vanish.

This pattern mirrors earlier automation waves more closely than popular narratives admit. Assembly lines reorganized factories around new roles. The Internet redefined clerical work. In each case, fear focused on displacement while reality delivered redistribution. The difference with humanoids is that embodiment makes the redistribution feel more personal.

That emotional response is understandable. When a machine shares your physical space and mirrors your movements, it feels like a competitor rather than a tool. The concern is not always articulated clearly, but it surfaces in questions about intent, autonomy, and scale. Addressing it requires acknowledging the emotional layer while examining the structural one.

My own conviction on this issue came from watching how fragile these systems still are, and how much human judgment is required to make them useful. Even as capabilities improve, the complexity of coordination grows alongside them. More capability creates more edge cases, not fewer. Humans remain essential precisely because the system is becoming more powerful.

What Operators Actually See

There is also a demographic reality that rarely enters public discussion. Many societies face labor shortages driven by aging populations, declining birth rates, and shifting preferences. In these contexts, humanoids are supplementing absent labor. The replacement narrative assumes a static labor pool that no longer exists in many regions.

In my conversations with operators and managers, the dominant concern is how to keep operations running when people are unavailable, unwilling, or physically strained. Humanoids are evaluated as pressure relief valves, not as substitutes. They are a way to maintain continuity, not to hollow out organizations.

Replacement would require not just better robots, but different laws, insurance models, labor norms, and cultural expectations. None of those are moving in lockstep.

Replacement fears are rarely about robots alone. They are about precarity, loss of control, and mistrust of institutions. Humanoids become a focal point for anxieties that predate them. Addressing those anxieties requires honesty about limits as well as possibilities.

From where I stand, humanoids are a response to work that has become misaligned with human well-being. They absorb tasks that humans should not be doing indefinitely. They force organizations to articulate what humans are actually good at, rather than assuming that all labor is interchangeable.

This rearticulation is uncomfortable. It shifts the conversation from replacement to redesign. The question stops being who disappears and becomes what changes.

Humanoids will change how work is structured, how value is created, and how roles evolve. That change will be uneven, contested, and slow.

The Work That Remains Human

If work is being reevaluated, the question becomes: What remains uniquely human when tasks shift?

Much of what people do each day does not appear clearly in process diagrams or operating manuals. It lives in judgment calls, informal coordination, and responsibility taken in moments when formal rules run out. These layers of work exist because real environments are complex, social, and imperfect. They persist precisely because no system, human or machine, can anticipate every condition in advance.

One of the first realizations that shaped my thinking is that human work is fundamentally contextual. People constantly interpret incomplete information and adapt their behavior based on signals that are hard to formalize. Tone of voice, hesitation, prior history, and unspoken expectations all influence decisions in ways that cannot be reduced to explicit rules without losing fidelity. When something unusual happens, people do not freeze waiting for instructions. They infer intent, assess risk, and act.

That capacity becomes more visible when automation increases. As soon as a system requires explicit boundaries, the surrounding human judgment stands out more clearly. Defining what a machine can and cannot do forces organizations to confront how much human discretion they have been relying on without acknowledging it.

Judgment, Responsibility, and Coordination

Human work requires responsibility under ambiguity. Humans are accountable not just for execution, but for outcomes in situations where the correct course of action is not obvious. When something goes wrong, people look to another person who can weigh intent, context, and consequence together. This expectation is deeply embedded in how trust operates. Even when systems assist with execution, responsibility remains human because it is tied to moral and social judgment, not mechanical output.

Organizations often underestimate how much of their stability depends on this kind of responsibility. It is rarely distributed evenly. A small number of individuals tend to carry a disproportionate share of interpretive load. These are the people others turn to when processes conflict, when exceptions arise, or when the cost of a decision extends beyond efficiency into safety, reputation, or fairness. These roles are rarely formalized, yet they are among the hardest to replace.

A large category of human work consists primarily of coordination. People spend significant energy aligning expectations, sequencing actions, and resolving misunderstandings between different parts of an organization. This work is often invisible because it does not produce a tangible artifact. Yet without it, systems fragment. Machines can execute defined actions efficiently, but they do not reconcile competing priorities or repair misalignment between groups.

Much of what people fear losing is not physical effort, but agency. Repetitive tasks may be tiring or even demeaning, but they are not usually where people anchor their identity. What people defend, often unconsciously, is the right to interpret, to decide, and to take responsibility for outcomes that affect others.

Tasks sit on top of judgment. Judgment sits on top of responsibility. Responsibility sits inside culture. Removing one layer does not eliminate the others. It changes how they are exercised and where they concentrate. When execution shifts, judgment and responsibility do not vanish. They become more central.

I stopped thinking about human work as something that needed to be defended from machines. Instead, I began to see it as something that needed to be articulated more clearly. Many organizations struggle not because they are over-automated, but because they have never explicitly named the human roles that keep them functioning. Advanced machines force that. They require organizations to say, out loud, what decisions matter and who owns them.

That process can be uncomfortable, but it is also constructive. It allows people to see their contribution more clearly and to develop skills aligned with judgment, coordination, and responsibility rather than endurance alone. When people understand how their work changes rather than disappears, fear loses some of its grip.

The work that remains human persists because it is structurally necessary. It exists at the intersection of uncertainty, accountability, and trust. As long as organizations operate in environments that are social, variable, and consequential, that intersection will require human presence. Machines can assist, amplify, and absorb certain forms of effort, but they do not dissolve the need for human judgment. They make it more explicit.

Cultural acceptance depends less on reassurance and more on whether people feel their role is becoming clearer rather than smaller. Humanoids do not erase the human role. They expose what it has been carrying all along.

And that exposure reveals the true constraint on this technology's future: not capability, not capital, but culture itself.

The Moment a Humanoid Enters Human Space

Arguments clarify. Presence transforms. Long before humanoids are deployed at scale, people already react to what they believe those machines represent. I encountered this repeatedly, in conversations, interviews, and closed-door discussions where the technology was still abstract. The moment that changed everything was when a humanoid stood in the room.

When humanoids are discussed on slides, debated in interviews, or imagined through videos, people project. They argue about replacement, timelines, and inevitability because nothing is yet anchored. Once a humanoid is physically present, people stop debating the future and start negotiating the moment. Where do I stand? How close is too close? Does it see me? Is it waiting? These reactions are responses to shared space.

I noticed early on that presence compresses conversation. In abstract discussions, people speak in long arcs about disruption and strategy. In rooms with a humanoid, language becomes more precise. Sentences shorten. Gestures become more deliberate. People point, pause, and confirm. This is not fear. This is orientation. The human nervous system is tuned to bodies, not ideas. A humanoid triggers that, even when everyone knows it is a machine.

This is where many misunderstand cultural resistance. Resistance is often framed as opposition to technology or anxiety about jobs. What I observed was something subtler and more grounded. People were responding to proximity. A humanoid introduces a new actor into the spatial and social contract, and contracts must be renegotiated before work can resume. That renegotiation happens instinctively, not ideologically.

In one early demonstration, the humanoid performed a simple, pre-scripted motion. Nothing complex. Nothing useful. Yet afterward, the conversation shifted toward responsibility and boundaries. Who is accountable if it moves differently next time? Who decides when it should stop? Who explains its behavior to others in the room? None of these questions had been raised in the planning meetings. They only surfaced once the machine stood there, quiet and patient, forcing the room to acknowledge it.

Humans are accustomed to sharing space with other humans whose intentions they can infer and whose behavior they can anticipate. Machines disrupt that pattern. Even when behavior is predictable, intention is opaque. Presence makes that opacity tangible.

I also noticed how quickly people assign character, even when told not to. A slight tilt of the head, a pause before motion, a change in stance, all invite interpretation. This is because social cognition is automatic. We read bodies before we read labels. A humanoid cannot avoid this. Presence makes neutrality impossible. The machine will be perceived as cautious, abrupt, hesitant, or confident whether designers intend it or not.

This is why proximity amplifies design choices that might otherwise seem cosmetic. Motion profiles, idle behavior, gaze direction, and recovery movements all take on cultural weight once the humanoid shares space with people. I have seen reactions swing dramatically based on small changes. A slower approach feels respectful. A sudden stop feels alarming. A brief pause before action feels thoughtful. None of these attributes improve task performance in isolation, but they profoundly affect acceptance. These effects appear within minutes, sometimes seconds.

What surprised me most was how presence reorganized group dynamics even in brief, non-operational settings. During public demonstrations, closed previews, and controlled walkthroughs, people often repositioned themselves unconsciously once a humanoid was physically present. They adjusted where they stood, how they moved around the space, and who spoke when. Informal roles emerged without being assigned, with one person instinctively explaining what was happening while another paid attention to distance or timing.

Humanoids provoke environments to organize around them. That organization can be constructive or chaotic depending on how the presence is handled. When presence is treated as an afterthought, confusion follows. When presence is acknowledged and shaped deliberately, people adapt quickly.

What Presence Demands

I learned to pay close attention to moments of silence. Silence around a humanoid is not absence of engagement. It is processing. People are recalibrating their internal models, watching for cues. They are deciding how much trust to extend and under what conditions. Rushing that process with explanations or promises often backfires.

This is also where the limits of technical language become clear. Engineers want to explain sensors, models, and safety cases. Those explanations matter later, but presence operates on a different register. People first want to know whether the machine respects space, whether it moves with awareness, and whether it will behave consistently.

I have seen situations where a humanoid with limited capability was received more positively than a more advanced system simply because its presence was calmer and more legible. Conversely, I have seen technically impressive systems rejected almost immediately because their movements felt abrupt or their pauses felt uncanny. Presence exposes mismatch faster than performance metrics ever could.

This is why cultural acceptance cannot be retrofitted. It must be designed from the moment a humanoid is meant to share space with people. That design means respecting the social contract that governs proximity, timing, and signaling. Presence without that respect creates friction long before any economic argument is made.

Another pattern became clear over time. Presence shifts accountability upward. When a humanoid is visible, responsibility cannot hide behind abstraction. Someone must answer for it. Someone must decide when it is appropriate to proceed, pause, or withdraw. This accountability pressure is healthy, but it can feel uncomfortable in organizations that are used to deferring responsibility to tools or procedures.

This pressure has surfaced in subtle ways. Managers ask more explicit questions. Safety officers request clearer boundaries. Operators seek confirmation before proceeding. These behaviors are often interpreted as resistance. In reality, they are signs that the system is being taken seriously.

How Fear Changes Form

Presence also reveals something else that surprised me. Many people are less concerned about replacement when the humanoid is actually there. Proximity replaces fear with assessment. People begin to ask what the machine can actually do, how reliably, and under what supervision. Those questions are grounded. They open the door to realistic roles rather than imagined threats.

This does not mean fear disappears. It becomes conditional rather than absolute. People worry about misuse rather than existence. They worry about pace rather than inevitability. Those concerns are addressable.

I have seen moments where someone who arrived skeptical left more open, not because the humanoid impressed them, but because it behaved within clear limits. The machine did not overreach. It waited. It deferred. It stopped when asked. Those moments matter more than flashy capability.

This is where emotional arc enters the story, not as sentimentality, but as pacing. Acceptance accumulates through repeated, uneventful encounters. A machine that does nothing wrong builds trust faster than one that does impressive things unpredictably.

I have learned to value boredom as a signal. When people stop paying attention to the humanoid because it has become predictable, a threshold has been crossed. Presence has normalized. Only then does it make sense to expand capability or autonomy. Before that point, more functionality often creates more anxiety. Progress in early phases is quieter integration.

Presence also exposes the importance of withdrawal. A humanoid that knows when not to act earns more trust than one that always tries. I have seen systems gain credibility simply by backing away when uncertainty increased. That behavior reads as respect. Presence includes the ability to reduce itself.

Ultimately, presence is where humanoids become real to people. Not useful yet. Not scalable yet. But real. That reality forces culture to engage. It surfaces fears, expectations, and values that no white paper can resolve. Ignoring that process does not make it disappear. It delays it until the stakes are higher.

The human element begins here, not at scale, not at replacement, but at proximity. Before humanoids change work, they change rooms. Before they alter economics, they alter behavior. Presence is the first inflection point, and it is irreversible once crossed.

Everything that follows—policy, economics, design, and deployment—is shaped by how that moment is handled. And once presence forces organizations to confront what humanoids actually are, something else begins to shift: how work itself is understood.

When Work Is Reevaluated

Once presence settles, the conversation changes. Public debates about humanoids tend to collapse quickly into outcomes: jobs lost, jobs created, entire categories rewritten overnight. But inside organizations, something quieter and more consequential begins: the reevaluation of work itself.

The moment organizations begin seriously evaluating humanoids, even in exploratory or hypothetical terms, the conversation is no longer about robots but about the work itself. Not future work. Present work. Specifically, work that has been held together for years by human accommodation rather than deliberate design.

Most organizations rely far more on implicit human behavior than they are willing to admit. People step in when processes are unclear. They absorb variability when systems fail. They stretch time, attention, and physical effort to keep operations running. It is culturally understood. It is how work gets done. When the possibility of a humanoid enters the discussion, that informal contract is the first thing that starts to strain.

The discomfort I've observed is rarely about replacement. It is about exposure. To even evaluate whether a humanoid could perform a task, teams are forced to describe what actually happens between instruction and outcome. They have to name tasks that were never formally named. They have to separate responsibility from availability, and effort from presence. In those moments, people realize how much of their value has lived in the gaps, not in the job description.

This realization lands unevenly. For some workers, it feels like overdue recognition. Work that was previously dismissed as routine suddenly reveals its complexity. For others, it feels like a loss of protection. Informal authority, built on being the person who "just knows how things work," becomes harder to defend once work is made explicit. That tension shows up early, often before any technical assessment has concluded.

On the management side, many organizations benefit from ambiguity, even if they don't acknowledge it. Vague expectations allow maximized output without confronting workload, pace, or exposure directly. When humanoids are evaluated seriously, those ambiguities become difficult to sustain. Systems that require explicit instruction do not tolerate implied effort. Conversations that were once framed as performance issues begin to surface as design choices.

What Exposure Reveals

Risk is another fault line that becomes visible almost immediately. Human labor has long served as a buffer against failure. When processes break, people compensate. When environments are unpredictable, people adapt. This risk absorption is culturally normalized and rarely accounted for. In pilot discussions and feasibility reviews, the question that stalls progress is often responsibility. Who has been absorbing risk until now, and what happens when that becomes explicit?

These conversations are uncomfortable because they imply moral responsibility, not just operational change. Safety stops being framed as individual resilience. Fatigue stops being treated as a personal shortcoming. Instead, both start to look like properties of the system itself.

Coordination exposes a similar pattern. Many workplaces function through tacit understanding. People know when to step in, when to wait, and when to bend rules. This works precisely because it is informal. The moment organizations begin specifying work clearly enough to evaluate humanoid use, that informality becomes a liability. Handoffs have to be defined. Escalation paths have to be named. Exceptions have to be handled deliberately. In every organization I've seen go through this exercise, that clarification redistributes power.

This is where resistance often emerges, even among teams that are otherwise aligned. There is resistance to losing discretion, status, or invisibility. Work that was once shielded by ambiguity becomes exposed to scrutiny. Decisions that were once absorbed quietly by individuals become shared responsibilities.

I've seen this trigger defensiveness more reliably than almost any other issue. In cultures that prize responsiveness, explicit time accounting feels like a threat. Workers worry about losing status. Managers worry about losing leverage.

Skill and status are renegotiated as well. Public narratives tend to frame automation as targeting low-skill work. In reality, the distinction that matters culturally is not skill level, but visibility. Work that is poorly specified, physically demanding, or socially invisible is often undervalued because it is taken for granted. When those tasks are examined closely enough to assess automation, their complexity becomes undeniable. At the same time, some roles lose mystique when their work is broken down into repeatable components. Authority based on opacity gives way to authority based on judgment and integration.

These shifts rarely produce hesitation. Meetings end with agreement but no decision. Pilots are approved but narrowly scoped. Progress becomes incremental not because capability is lacking, but because legitimacy has not yet been negotiated.

What Remains Human

Work does not lose its role as a source of identity or meaning. People do not become unnecessary because certain tasks can be specified or assisted. What changes is how human energy is justified and allocated. The possibility of reducing certain kinds of physical or cognitive load forces a renegotiation of what is reasonable to expect from people. That renegotiation is cultural before it is technical.

The deeper shift is from implicit labor to explicit systems. From unspoken sacrifice to designed responsibility. From norms that reward endurance to norms that demand accountability.

From where I sit, having watched these dynamics play out repeatedly across sectors and institutions, the most striking thing is how early they appear. They do not wait for deployment. They emerge the moment work is examined closely enough that human flexibility can no longer be treated as a free resource.

The more consequential change is already underway, and it is happening in language, planning, and expectation. Organizations that engage honestly with this process tend to become more deliberate about how they structure work, whether or not a humanoid is ever introduced.

What the consideration of humanoids introduces, culturally, is a present in which assumptions about work are harder to avoid. Cultural acceptance will depend less on proving what humanoids can do, and more on how honestly societies are willing to confront what they have been asking people to do all along.

Culture Is the Real Constraint

Culture became unavoidable because partial capabilities entered real conversations without a corresponding agreement on permission, responsibility, or normal behavior. This is the phase where systems stop being theoretical and start demanding decisions that no one is structurally prepared to own. Humanoids now sit in that space, not deployed at scale, but present enough to force alignment questions that institutions, markets, and societies have not yet resolved.

Early on, culture was not the limiting factor. There was nothing to negotiate. Humanoids were brittle, narrow, and clearly experimental. They lived in labs, research programs, and demonstrations that carried no expectation of operational continuity. Institutions simply did not take them seriously. Culture does not block what it does not recognize as plausible.

That changes in the middle zone. This is the phase where systems are good enough to matter but not complete enough to settle uncertainty. Humanoids can perform tasks in constrained settings. They can operate under supervision. They can appear in pilots, briefings, and walkthroughs without collapsing. At that point, cultural forces begin shaping outcomes, not as opposition, but as hesitation. The question shifts from whether this could work to who is allowed to say yes, and under what conditions.

Inside organizations, this hesitation shows up as agreement without motion. Leaders acknowledge that humanoids are strategically relevant. Operators see potential operational value. Innovation teams advocate for experimentation. Yet decisions stall at the moment where endorsement turns into ownership. Procurement cannot classify the system cleanly within existing categories. Legal teams cannot anchor liability to established precedent. Risk functions cannot quantify exposure because behavior is adaptive rather than fixed. Each group is acting rationally within its mandate, but together they produce inertia.

What is often described as caution is more accurately a lack of cultural authority. No role inside the organization is clearly empowered to define what acceptable use looks like for an embodied system that operates with partial autonomy. Without that authority, every decision feels provisional. Provisional decisions rarely survive escalation, because escalation demands justification, and justification demands norms that do not yet exist.

This dynamic becomes even clearer in pilots and demonstrations. Scope narrows because expanding the scope expands accountability. Timelines stretch because each successful milestone introduces new questions that cannot be resolved technically. Who intervenes if behavior deviates? Who explains outcomes to internal stakeholders? Who absorbs blame if something goes wrong, even when no rules were broken? These questions return repeatedly because they are social questions framed as technical ones.

As performance improves, these tensions intensify rather than resolve. Success removes the excuse of immaturity without providing a replacement framework for responsibility. The system is no longer safely hypothetical, but it is not yet culturally normal. Pilots linger in a state of permanent rehearsal, demonstrating capability without authorizing continuation. Culture asserts itself precisely at this point, not to stop progress, but to slow it until interpretation catches up.

This is where culture becomes visible as a constraint. It has not yet produced shared norms for living with it. Culture is the process through which societies decide what is ordinary, what is tolerated, and what requires oversight. That process cannot be completed in advance. It only begins once systems are present enough to demand judgment, and judgment requires context, precedent, and shared understanding.

Accountability sits at the center of this tension. Humanoids strain existing assumptions about ownership because they do not fit neatly into established categories. They are not tools in the traditional sense, and they are not agents in the human sense. Responsibility diffuses upward because no one wants to be the first to normalize a new class of behavior without precedent. The perceived cost of being wrong feels higher than the cost of waiting, even when waiting carries its own risks.

This pattern is familiar from autonomous vehicles. Capability improved steadily. Pilots expanded. Public belief oscillated. Regulation evolved incrementally. Cultural permission, however, never arrived cleanly. Partial deployment became a stable condition rather than a stepping stone. Humanoids differ in form and context, but the structural tension is the same. Visibility accelerates cultural negotiation without resolving it.

Language plays a critical role in this phase. Certain framings tighten rooms even when there is broad agreement on long-term value. Claims that imply inevitability or replacement trigger defensive responses because they sound like loss of control. Framing humanoids as bounded participants rather than autonomous actors reduces resistance, not because it changes capability, but because it aligns with how institutions manage uncertainty. Culture responds to framing long before it responds to performance metrics or benchmark data.

Media narratives and historical associations amplify these dynamics, raising the perceived stakes of each decision.

Importantly, culture does not relax once humanoids appear in limited deployments. It intensifies. Early presence forces decisions that could previously be deferred. Visibility, behavior norms, escalation paths, and explanation mechanisms all become operational concerns. Trust becomes harder to calibrate, not easier, because stakeholders must decide how much reliance is appropriate without precedent. In the absence of shared norms, caution becomes the default.

Labor identity adds another layer of complexity. Work is not only economic; it is cultural. Humanoids intersect with labor not just by performing tasks, but by challenging how contribution, supervision, and responsibility are defined. Even when framed as assistance, their presence raises questions about role boundaries and professional meaning. Resistance here is about preserving human agency within systems that are increasingly mediated by automation.

This implies negotiation. Societies do not absorb new categories of capability smoothly, especially when those capabilities are embodied and adaptive. Culture moves unevenly across organizations, sectors, and regions. The same system can be interpreted differently depending on institutional norms, labor structures, and social expectations. This variation does not signal confusion. It signals that norms are still forming under real conditions.

The real risk is misreading cultural hesitation as something to bypass rather than something to engage. When cultural questions are ignored, they harden into policy, stigma, or frozen assumptions that are difficult to unwind later. When they are acknowledged early, they can inform design choices, deployment strategies, and governance models that support gradual normalization without forcing premature resolution.

Culture is now the real constraint because multiple systems are advancing without synchrony. Capability, belief, permission, and responsibility are all in motion, but none are aligned. Humanoids surface this misalignment earlier than most technologies because they cannot be abstracted away. They demand interpretation simply by being present, even in constrained and provisional settings.

This does not mean progress is stalled. It means progress is negotiating its own legitimacy. That negotiation will evolve as humanoids move from demonstrations to pilots to constrained operations, and as societies decide, context by context, what role these systems are allowed to play.

Acceptance Is Never Settled

Culture does not stop constraining humanoids once initial permission is granted; in many ways, it becomes more active. Early acceptance is often conditional, bound to a task, a site, or a moment in time. That conditionality is not a weakness. It is how culture tests unfamiliar systems.

As humanoids move from demonstrations into limited operations, the nature of cultural scrutiny changes. Early questions focus on whether the system should exist at all. Later questions focus on how far its role is allowed to extend. A task that is acceptable in one context may feel inappropriate in another, even if the technical behavior is identical. Culture renegotiates permission locally.

This is why progress often appears uneven. A humanoid may be welcomed in a warehouse aisle but questioned in an adjacent break area. It may be accepted during peak demand but resisted during normal operations. These shifts reflect how humans draw boundaries around spaces, moments, and roles. Cultural acceptance is granular by default, even when technology aspires to generality.

Organizations experience this as friction during expansion. A pilot that succeeded quietly in one unit encounters resistance when replicated elsewhere. New managers ask questions that were already answered. Safety teams request clarifications that already exist. Workers compare the system to their own norms rather than to prior deployments. From the outside, this looks like regression. From the inside, it is culture doing its work again because the context has changed.

This persistence of negotiation is often underestimated. Many teams treat cultural acceptance as a gate to pass through, after which the scale becomes a technical exercise. In practice, every new task, environment, or interaction reopens the question of fit. Culture responds to lived experience, not roadmaps. It reacts to how the system behaves in a specific place, with specific people, under specific pressures.

Public perception follows a similar pattern. Initial curiosity fades, but scrutiny does not disappear. Instead, it shifts toward expectations. Once a humanoid is no longer novel, people judge it against norms of reliability, courtesy, and restraint. Small deviations that would have been excused early on become more salient. Trust becomes conditional on consistency rather than possibility.

Familiarity raises the bar. Once people integrate a system into their mental model of the environment, they expect it to behave like other stable elements of that environment. When it does not, cultural tolerance tightens. Acceptance matures into expectation, and expectation brings pressure.

What Culture Remembers

There is also a temporal aspect that complicates planning. Cultural memory accumulates. Early incidents, even minor ones, shape long-term attitudes. Successes fade faster than failures. A single unresolved moment can influence how future proposals are received, even if the system improves technically. Culture remembers through stories, not metrics, and those stories travel unevenly across teams and institutions.

This is why culture evolves alongside deployment. Each expansion reactivates questions of responsibility, boundaries, and legitimacy. Each new context forces reinterpretation. Progress depends not on eliminating this process, but on designing deployments that can withstand repeated scrutiny without escalation.

Seen this way, cultural constraint is not an external drag on innovation. It is an internal pressure that shapes how systems earn durability. Humanoids that succeed are not those that move fastest, but those that survive repeated renegotiation without provoking backlash. They remain legible, bounded, and predictable even as their role evolves.

Culture remains the constraint because legitimacy is cumulative and fragile. Each step forward carries memory. Each new use case reopens judgment. Humanoids will continue to operate under this condition for as long as they share space with people. Recognizing that reality does not slow progress. It makes progress survivable.

7

The Global Stakes

CULTURE DOES NOT stop at organizational boundaries. The same forces that shape acceptance inside a company—legitimacy, memory, hesitation, negotiation—operate at the level of nations and institutions.

Humanoids have become an international phenomenon. Across the world, people are interacting with robotics, refining designs, and building on each other's work. The movement now reaches small innovators and large technology companies alike. Progress is no longer confined to a few labs or a single region. It is distributed and accelerating.

That distribution changes the nature of the challenge. The questions raised within organizations about labor, autonomy, accountability, and trust are now surfacing between nations. The actors shift from operators and managers to institutions, alliances, and competing visions of how embodied intelligence should develop. The stakes shift from whether a pilot succeeds to whether an ecosystem coheres.

What happens when cultural constraint meets geopolitical reality? How international communities respond to humanoids will shape more than markets. It will shape what kind of future becomes possible. The question is no longer just how humanoids earn trust. It is who shapes the terms on which that trust is offered.

Why Humanoids Become Strategic National Assets

The moment humanoids became strategic was not announced. It happened quietly, in rooms where the questions changed. I have seen this shift multiple times, across different institutions, and it always follows the same pattern. Early conversations focus on capability and timelines. Later ones focus on continuity, failure tolerance, and what happens when human labor is no longer elastic. Once that transition occurs, humanoids stop being discussed as technology and start being discussed as infrastructure.

What triggers the shift is rarely a breakthrough. It is usually a constraint: Labor gaps that do not close, maintenance backlogs that grow despite hiring, sites that operate close to failure margins because experienced staff are aging out faster than they can be replaced. In those settings, the question is no longer whether automation improves efficiency. It is whether the system can hold under pressure. That is when humanoids enter the frame as a strategic consideration, even if no one believes they are ready for broad deployment.

Slack has become taboo. For years, efficiency was treated as a virtue. Headcount was trimmed. Margins were tightened. Redundancy was framed as waste. Systems learned to operate at their limits because they were rewarded for doing so. When everything worked, the model looked strong. When conditions shifted, there was nothing left to absorb shock.

Humanoids enter this conversation as a way to reintroduce margin without reopening every labor contract or redesigning every facility. They are considered because they can occupy human-scale work without demanding human-scale fragility. They do not get sick. They do not age out of physically demanding roles. They do not require decades of tacit knowledge to be productive in bounded tasks. In environments where experience is thinning and recovery time matters, that combination carries weight.

I have watched seasoned operators explain this more clearly than any strategist ever could. They talk about breathing room. About having something that can cover the third shift when staffing collapses. About reducing the number of moments where a single injury or absence cascades into missed deadlines, unsafe shortcuts, or forced overtime. In those explanations, humanoids are never framed as the center of the system. They are framed as load-bearing relief.

This is why strategic interest often appears first in maintenance, logistics, and support roles rather than in core production. These functions sit at the edge of failure. They are essential, but chronically understaffed and under-modeled. They rely on human adaptability to compensate for brittle processes. When adaptability becomes scarce, leaders look for capacity that does not require rewriting the environment. Humanoids are evaluated because they fit into the existing geometry of work.

That fit matters more than capability curves. A system that performs slightly fewer tasks but can be deployed without retooling an entire site will be taken seriously faster than a more capable system that demands structural change. Strategic planning favors integration over elegance. It rewards what can be absorbed under pressure, not what looks impressive in isolation.

Instead, the questions shift quietly. People stop asking how soon and start asking how safely. They stop asking what it can do and start asking what it cannot be allowed to do. They stop tolerating ambiguity, not because they are hostile, but because the consequences of ambiguity have become real.

Once that happens, the label follows naturally. Anything that touches continuity, safety, and recovery becomes strategic, regardless of its original intent. Humanoids fall into that gravity well because they intersect with labor, space, and time simultaneously. They do not need to be widespread to matter. They only need to be plausible enough to be considered in contingency planning.

What Strategic Attention Brings

Once a capability is treated as strategic, a second-order shift follows. Questions about origin, supply chain, and control surface immediately. Conversations pivot from pilot design to component traceability in a single meeting. Governments want to know where systems are built, how they are serviced, and which dependencies could become bottlenecks. This is not ideological. It is operational risk management.

Strategic assets are expected to be dependable under constraint, which means their inputs and update paths matter as much as their performance.

This is where the framing can easily go wrong. Strategic importance is often conflated with militarization or secrecy. In practice, most of the strategic value in humanoids is civilian. It comes from safer industrial work, steadier logistics, better maintenance coverage, and faster recovery after disruption. When those benefits are visible, the strategic argument does not need to be amplified. It becomes self-evident. When the category is framed narrowly around defense, public trust erodes and the ecosystem narrows prematurely.

Capital behavior changes once the strategic label appears. Funding shifts from exploratory to programmatic. Acceptance criteria replace demos. Service, training, and reporting matter more than novelty. This raises expectations across the stack. Teams are asked to operate with transparency, to document behavior, and to support long-term operations rather than short pilots. For builders who can meet that bar, the strategic turn brings clarity and stability. For those who cannot, it exposes gaps that were previously hidden by excitement.

There is a real risk of capture in this phase. Strategic customers bring scale and credibility, but they also shape road maps. Product decisions begin to align with funded deployments. Compliance requirements harden. Export rules and reporting obligations influence architecture. The challenge is not whether to engage, but how. Products that are designed to serve only one strategic customer or one national context lose flexibility and long-term value. Products that can satisfy strategic requirements while remaining legible to civilian operators retain optionality.

From an ecosystem standpoint, the strategic turn has cascading effects. Suppliers invest in certification and traceability. Integrators build service models that assume fewer specialists on site. Workforce programs adapt to include supervision and recovery roles. Insurers begin to price risk based on evidence rather than assumptions. Each of these developments reduces friction for future deployments, including those that are not strategically funded. Strategic attention, when handled well, lifts the entire field.

The Risk of Fragmentation

The alternative is fragmentation. When every country treats humanoids as a closed national project, costs rise and learning slows. Standards diverge. Interfaces fracture. Smaller players are excluded. I have seen how quickly this can happen when strategic framing turns inward. Resilience through isolation feels appealing, but it produces brittle systems. Resilience through redundancy and shared proof is harder to coordinate, but it scales better and lasts longer.

What matters in this phase is not speed for its own sake, but seriousness. Serious safety cases. Serious service models. Serious workforce plans. Serious evidence. When those are present, governments can support the category without freezing it into a single path. When they are absent, strategic interest turns into control, and progress stalls under its own weight.

Strategic importance raises the stakes. It also clarifies priorities. Reliability over spectacle. Proof over promise. Integration over isolation. Those choices will determine which ecosystems endure and which ones collapse under their own constraints.

How Strategic Attention Reshapes the Ecosystem

The strategic label arrives as pressure. Once humanoids are discussed as something that could matter to continuity rather than innovation, the ecosystem around them begins to change shape. The shift is subtle at first, but it compounds quickly. Conversations that once tolerated ambiguity start demanding definition. Systems that were allowed to be impressive are now expected to be dependable.

Exploratory interest gives way to conditional seriousness. Meetings that once focused on demonstrations start centering on acceptance criteria, service assumptions, and failure handling. The questions narrow. How does this system behave after twelve hours, not twelve minutes? What does recovery look like when a site is short on staff? How much training is required before supervision is competent rather than nominal? These are the questions of organizations that expect to live with a system.

Procurement reflects this shift immediately. Early pilots are often loose by necessity, designed to learn quickly rather than to commit.

Once a capability is treated as strategic, that looseness disappears. Programs harden. Scope is bounded. Safety envelopes are declared. Reporting requirements appear. Multi-year thinking replaces short cycles. This compresses timelines in one sense, because decisions are made with intent. It also raises the bar, because incomplete answers are no longer acceptable. Teams that cannot operate under audit pressure fall away.

Inside companies, road maps begin to bend toward funded use cases. Test suites prioritize environments that mirror strategic sites. Supplier choices shift toward components with certifications, traceability, and predictable availability. Documentation grows heavier. None of this is inherently negative. It is what happens when customers face consequences for failure. The pressure is constructive when it forces discipline.

Capital behavior changes in parallel. Strategic interest alters how risk is underwritten. Investors begin to care less about novelty curves and more about durability. Service models, workforce training, and compliance readiness carry real weight. This favors teams that can show operational maturity, even if their raw capability advances more slowly.

It also creates a bifurcation. Some builders optimize for strategic programs early and accept the constraints that come with them. Others stay intentionally civilian and risk slower uptake. Both paths can work. The danger lies in drifting between them without choosing.

Standards, Evidence, and the New Currency of Trust

Standards become leverage points once strategy enters the picture. Early technical agreement is usually easy. Engineers often align on safety logic, test methods, and basic behaviors. Tension appears when standards are asked to serve national or institutional goals. Small clauses begin to matter. Definitions stretch. Conformance becomes a gate rather than a guide. The antidote is not neutrality, which is rarely possible. It is testability. When standards are anchored to observable behavior and reproducible tests, there is less room for interpretation to hide incentives.

Inspectability becomes the currency of trust. Logs, proofs, and artifacts matter more than claims. Systems that can show what happened, when, and under what constraints move faster through review. Systems that rely on explanation rather than evidence slow down, regardless of their capacity. This is not because reviewers distrust builders. It is because responsibility has shifted. When outcomes matter beyond a single site, evidence replaces narrative.

Supply chains reshape accordingly. Once a capability is considered strategic, questions of origin and dependency surface immediately. Officials want to know which parts are critical, where they come from, and how easily they can be substituted. Incentives appear to localize certain components. Export controls tighten around others. Done carefully, this improves resilience and traceability. Done aggressively, it fragments markets and raises costs without improving reliability. The practical line is redundancy over isolation. Multiple qualified sources. Clear interfaces. Conformance tests that allow parts and assemblies to move without bespoke negotiation.

Dual-use is not an exception in this environment. It is the default condition. The same actions that move industrial goods can move emergency supplies or maintenance equipment. Treating this reality as a paperwork problem rather than a design constraint creates friction later. Teams that plan for variant control, scope limitation, and precise logging avoid expensive rewrites. Teams that defer those questions often find themselves blocked by reviews that arrive too late to accommodate change.

Workforce implications sharpen as well. Strategic programs bring funding for training and supervision, but they also raise expectations. Operators are no longer passive recipients of a tool. They are part of a system that must hold under stress. This elevates roles that were previously informal. Supervision, recovery, and coordination become explicit responsibilities. When this is acknowledged and supported, programs expand. When it is ignored, resistance grows quietly and surfaces later as attrition or pushback.

Insurance and liability follow the same pattern. Early on, risk is priced from analogy and fear. As programs mature, evidence begins to matter. Consistent incident taxonomies, clear autonomy bounds, and immutable logs give insurers something to work with. Premiums respond. Approval cycles shorten. This feedback loop is slow, but it is real.

The Cost of Closing Doors

There is a less visible cost to all of this. As programs harden, openness can suffer. Sensitive details retreat behind classification or contract walls. Lessons learned are not shared. The field relearns the same basics in parallel. Strategic seriousness does not require opacity, but it often produces it by default. The teams that resist this tendency, by publishing artifacts and methods where possible, tend to shape the ecosystem more than their market share alone would suggest.

Strategy changes the rules of engagement before it changes outcomes. It alters who is allowed to participate, how decisions are justified, and which behaviors are rewarded. Some teams thrive under that shift. Others retreat. The difference is rarely technical capability alone. It is whether the organization can operate transparently, accept constraints, and still preserve enough flexibility to evolve.

This is why the strategic turn is both an opportunity and a filter. It brings resources, clarity, and urgency. It also narrows tolerance for ambiguity and improvisation. Ecosystems that adapt consciously gain durability. Ecosystems that react defensively fragment. That is where coordination either emerges or fails, and where the global stakes become unavoidable.

Fragmentation or Coordination

Once humanoids are treated as strategic, coordination stops being optional. It becomes a pressure point. Strategy concentrates decision-making locally, but it exposes friction globally. The same forces that push institutions to take humanoids seriously also pull ecosystems apart. This tension rarely announces itself as conflict. It shows up first as divergence.

Interfaces drift. Safety cases stop lining up. Data expectations change quietly. Over time, those small differences accumulate into real barriers. What began as a reasonable local adaptation hardens into incompatibility.

Early alignment is easy when nothing is operational, and the stakes are low. As soon as systems move from pilots into sustained use, institutional defaults assert themselves. Agencies optimize for their own risk frameworks. Operators tune systems to their own environments.

Vendors adapt to the loudest customer in the room. Together, they fragment the field.

The Cost of Divergence

The cost of fragmentation appears as duplicated work, slower learning, and rising integration overhead. Teams rebuild the same safety logic to satisfy slightly different interpretations. Logs are reformatted to fit local review tools. Interfaces are rewritten site by site. Smaller builders struggle to keep up, not because they lack capability, but because the connective tissue keeps shifting beneath them.

In the short term, large players absorb these costs. In the long term, everyone pays them.

Coordination, when it works, relocates competition. When basic behaviors, safety proofs, and audit formats align, builders are free to compete on performance, reliability, and service. When those foundations diverge, competition collapses into compliance. Innovation slows not because ideas dry up, but because effort is consumed by reconciliation.

This is where strategic framing cuts both ways. Governments want assurance that systems behave predictably within their jurisdictions. That desire is legitimate. The failure mode comes when assurance is pursued through bespoke requirements that cannot travel. A rule that is clear but incompatible becomes a wall. A test that proves safety in one place but cannot be reproduced elsewhere becomes a bottleneck. Over time, these choices shape markets more than technology does.

Coordination succeeds when it is scoped narrowly. Attempts to align entire stacks almost always fail. Incentives are too strong. Contexts are too different. Alignment works better at the level of observable behavior. What constitutes a safe stop? How a near miss is recorded. What evidence is required to show a task stayed within bounds? These questions are practical and testable. Agreement here reduces friction without forcing convergence elsewhere.

The Narrow Window for Alignment

There is also a timing problem that is often underestimated. Align too early, and you lock in assumptions that practice has not yet validated.

Align too late, and fragmentation has already set. The window is narrow, and moves as capability matures.

The incentive to fragment is strongest when programs are under pressure. When resilience becomes urgent, institutions favor control over compatibility. They optimize for what they can guarantee locally. That instinct is understandable. It is also self-limiting. Systems built to operate only within narrow bounds struggle to adapt when conditions change. Resilience achieved through isolation often fails at the first unexpected interaction.

Trust requires evidence, and evidence requires shared proofs. When teams can point to the same tests, the same logs, and the same failure taxonomies, disagreement becomes manageable. When everything is bespoke, disagreement becomes structural. The difference is operational.

What is at stake here is legibility. A global ecosystem does not need identical systems. It needs systems that can understand each other well enough to coexist. Legibility lowers the cost of entry, speeds learning, and keeps strategic pressure from collapsing the field into a handful of closed stacks. It also gives operators and regulators confidence that they are not committing to dead ends.

This tension is shaped by choices made incrementally, often without fanfare. A standard adopted here. An interface closed there. A reporting format published or withheld. Over time, those choices determine whether the field remains navigable or splinters into silos.

The next layer of complexity appears when regional differences enter the picture as defaults.

Regional Gravity Without National Stereotypes

Regional differences matter in this field, but not in the way they are usually discussed. The most consequential distinctions are about how institutions behave under uncertainty, how risk is interpreted, and how legitimacy is earned. Those patterns shape adoption more than technical capability does, and they are remarkably consistent once you start watching for them.

Regions carry defaults. These defaults are rarely written down in one place. They show up in how meetings are run, how pilots are framed, and how long it takes for a tentative yes to turn into a real commitment.

They shape what kinds of evidence are persuasive and which kinds of arguments fall flat. Over time, they become gravitational forces that pull programs in particular directions, often without anyone naming them explicitly.

Four Patterns of Regional Default

Some ecosystems privilege speed and optionality. In these environments, early pilots are easier to launch and capital moves quickly. Teams are encouraged to learn in public, even if that means revising plans frequently. The upside is momentum. Ideas get tested, discarded, and refined rapidly. The downside is churn. Suppliers change, priorities shift, and programs can reset just as they begin to stabilize. For humanoids, this favors builders who can iterate fast without losing coherence.

Other regions privilege stability and proof. Decisions take longer, but once made, they tend to hold. Safety cases are expected to be thorough. Documentation is taken seriously. Operators are trained carefully before systems touch real work. The entry cost is higher, but the payoff is durability. Programs that clear these hurdles often scale more steadily because legitimacy has been earned early. In these settings, humanoids are evaluated as obligations that must behave predictably under scrutiny.

There are regions where coordination is the dominant feature. Industrial policy, supplier networks, and local authorities align quickly once a direction is chosen. Tooling, manufacturing, and deployment can move from concept to execution with impressive speed. The tradeoff is constraint. Strategic priorities narrow what can be shared, exported, or discussed openly. Builders operating in these contexts gain scale, but they also inherit policy complexity that must be managed continuously.

In other environments, regulatory depth shapes the field. Labor protections, privacy expectations, and liability frameworks are deeply embedded. Approval processes can feel heavy, especially to teams used to looser norms. The benefit is predictability. Once a path is defined, operators and insurers have confidence. Programs that pass serious review are more likely to be accepted by workers and communities. For humanoids, this can slow early progress but smooth later expansion.

These patterns overlap in practice. What matters is how they interact with a technology that sits at the intersection of labor, space, and autonomy. Humanoids are unusual in that they surface cultural and institutional assumptions immediately. The same pilot can feel straightforward in one region and deeply problematic in another, even when the technical behavior is identical.

Teams stumble by treating these differences as noise. They assume that success in one context will translate directly to another. When it does not, they blame regulators, culture, or politics. More often, the issue is misalignment. Evidence was presented in a form that did not carry weight locally. Risk was framed in a way that did not match institutional memory. Stakeholders who expected to be consulted were bypassed, and trust was lost early.

Navigating Without Chasing Uniformity

Teams that navigate this well maintain a stable core and adapt at the edges. Safety logic remains consistent. Logs remain legible. Task boundaries remain clear. What changes is how those elements are introduced, reviewed, and explained.

The same applies to workforce dynamics. In some regions, workers expect to be involved early and trained deeply before any system is introduced. In others, acceptance grows through exposure and iteration. Neither approach is inherently superior. What matters is matching deployment posture to expectations. Programs fail when workers feel surprised or sidelined. They succeed when people understand what is changing and why.

Privacy and data expectations vary in similarly structured ways. Some environments are comfortable with localized sensing as long as retention is minimal and purpose is clear. Others require formal consent processes and strict limits on use. Products that cannot accommodate these differences face long delays or rejection. Products that make data handling visible and configurable move more smoothly across borders.

Capital responds to these defaults as well. Investors underwrite timelines differently depending on where a company plans to deploy first. A slow, proof-heavy market may look unattractive if expectations are misread. A fast, fluid market may look deceptively easy until churn sets in.

Regional gravity shapes design. Builders internalize expectations from their first serious customers. A team that grows up in a proof-heavy environment will build differently from one shaped by rapid pilots. Neither approach is wrong, but they are not interchangeable. When teams cross borders, those inherited assumptions become visible, sometimes uncomfortably so.

Regional diversity is not an obstacle to overcome; it is a condition to work within. Humanoids are not neutral objects dropped into a vacuum. They enter systems with memory, norms, and power structures. Treating those systems as interchangeable leads to friction that no amount of technical progress can erase.

The global stakes are decided by whether regions can learn from each other without forcing convergence where it does not belong. That balance is fragile. It is shaped by choices that builders, institutions, and investors make long before scale appears.

Once those forces are understood, governance stops looking like bureaucracy and starts looking like infrastructure. Not something imposed from above, but something that allows systems shaped by different defaults to operate in the same world without breaking trust.

Governance Becomes Infrastructure

Governance enters the humanoids conversation the same way strategy does: as a response to pressure. As long as humanoids are treated as experimental, governance can remain informal. Teams rely on internal judgment, ad hoc reviews, and goodwill between partners. Once systems are expected to hold under real conditions, that informality breaks down. Decisions need structure, and structure needs to be shared.

Governance has been blamed for slowing progress, when the real issue was that it had never been built into the system in the first place.

The programs that move fastest treat governance as part of the technical stack. Safety constraints are explicit, not implied. Behavior is logged in ways that can be audited without interpretation. This does not eliminate review. It shortens it. When evidence is legible, questions close faster. When evidence is ambiguous, everything slows.

Logs are the clearest example. Early pilots often treat logging as an internal diagnostic tool. Once governance matters, logs become the record of truth. They establish what the system did, under what conditions, and who had control at each moment. When they could be read by someone outside the vendor's engineering team, trust formed quickly. When they could not, confidence eroded regardless of capability.

Standards play a similar role. At their strongest, they are shared test methods. A safe stop that can be demonstrated under defined conditions carries more weight than pages of narrative about safety philosophy. A duty-cycle report that any operator can generate matters more than a polished case study. Governance accelerates when standards describe behaviors that can be observed, repeated, and verified.

Many governance efforts fail because they try to settle questions that practice has not yet clarified. Committees attempt to define entire architectures or usage models too early. The result is delay, disagreement, and fragmentation. The governance that holds has a narrow scope. It focuses on outcomes that matter across contexts. What happens when a human enters a defined zone? How are exceptions handled? How is recovery initiated?

Where Governance Becomes Unavoidable

Liability is where governance becomes unavoidable. As long as responsibility is abstract, everyone agrees. When something goes wrong, abstraction collapses. The workable arrangements tie responsibility to control. When a vendor controls a behavior, the vendor is responsible for it. When an operator overrides constraints or sets policy, the operator is responsible for it. Immutable logs become the tie-breaker. Without them, disputes become political. With them, they become procedural.

Insurance follows the same logic. Consistent incident taxonomies, clear autonomy bounds, and documented recovery paths give insurers something concrete to work with. Programs that produce clear evidence find coverage earlier. Programs that rely on assurances wait longer, regardless of technical maturity.

Governance also shapes workforce acceptance. Operators are more willing to work with systems that have clear escalation paths and documented limits. Supervisors are more confident when responsibility is defined rather than implied. Training becomes meaningful when it aligns with real procedures and real constraints. When governance is absent, workers are asked to absorb risk quietly. When it is present, risk is managed openly.

Transparency in this context is not a moral posture. It is a functional one. Programs that publish safety case outlines, incident categories, and review outcomes move faster because fewer assumptions are made about their intent.

The Infrastructure That Emerges

Funding plays an underappreciated role here. Bodies expected to set norms without resources move slowly. Testbeds that depend entirely on sponsorship tilt toward whoever pays. The governance that holds has modest but stable support and a clear mandate. It produces artifacts on a cadence. It publishes drafts. It incorporates feedback. It stays close to practice. When governance drifts into symbolism, trust fades.

Timing is the hardest variable. Freeze rules too early, and you lock in fragile assumptions. Wait too long and fragmentation sets in. Safety case structure, incident taxonomy, and basic duty-cycle reporting tend to converge early. Manipulation strategies, planning architectures, and higher-level autonomy behaviors should not.

Over time, what emerges is not a single rulebook, but a mesh. Laws define outer limits. Standards define testable references. Review bodies translate those references into approval processes. Operators apply them in context. None of this works without feedback. When practice exposes a gap, the mesh adjusts. When governance is treated as static, it breaks under pressure.

The most important shift is conceptual. Governance becomes something the system produces. Logs, proofs, and constraints are not compliance artifacts. They are how the system earns the right to operate. Teams that understand this build differently. They design for inspection. They assume scrutiny. They make it easy to answer hard questions without theater.

This has implications for how builders organize internally. Teams that separate engineering from compliance too sharply create friction later. Governance decisions made downstream force architectural compromises upstream. The builders who move cleanly through review tend to integrate these concerns early, not by slowing development, but by making constraints explicit and configurable from the start.

As humanoids move closer to sustained use, governance will not feel optional. It will feel like infrastructure. Invisible when it works. Impossible to ignore when it fails.

The remaining question is not whether governance will exist. It is how it will be exercised. Whether it becomes a blunt instrument that narrows the field or a shared substrate that allows progress without sacrificing trust. That question is answered in how systems behave, how evidence is shared, and how responsibility is handled when things do not go as planned.

What remains are choices about boundaries, use cases, and restraint that cannot be fully automated or standardized. This is where the global stakes shift again, from structure into responsibility, and where the final layer of this chapter takes shape.

Choosing the Shape of What Comes Next

By the time a technology becomes strategic, momentum is already present. Decisions begin to compound faster than narratives can keep up. That is where humanoids now sit, as a force that is starting to pull labor, capital, policy, and culture into the same orbit.

What happens next will not be decided by a single breakthrough, a single company, or a single country. It will be decided by thousands of choices that look small in isolation. How a task is defined. How a safety limit is enforced. How a system behaves when conditions drift from plan. How responsibility is assigned when something goes wrong. These choices are already being made quietly, inside procurement language, test procedures, standards drafts, board discussions, and on real work sites.

That is how direction sets in a field like this: through defaults. Geopolitics will shape those defaults whether anyone intends it or not. The global landscape is not a single chessboard with one winner. It is a set of overlapping constraint systems. The United States optimizes for speed and capital formation. Parts of Asia optimize for coordination and execution. Europe optimizes for legitimacy, labor protection, and formal accountability. None of these approaches is complete on its own. Each solves for a different failure mode. The risk is divergence without connection.

A world of incompatible humanoid ecosystems would slow learning, raise costs, and concentrate power in ways that invite backlash. A world of compatible but competitive ecosystems can do the opposite. It can allow progress to travel while values remain local. It can let safety logic, auditability, and accountability form a shared floor, while implementation reflects regional priorities. That balance does not happen automatically. It has to be designed for.

Where Responsibility Becomes Operational

Moral responsibility in this context is operational. Capability alone does not determine impact. How a system is governed, constrained, explained, and accepted matters more than what it can do on paper.

Responsibility shows up in the constraints a system respects when no one is watching. It shows up in how data is handled, how consent is earned, and how incidents are reported. It shows up in whether workers are treated as partners in deployment or as variables to be managed around. Responsibility here is all about structure.

Humanoids will amplify whatever posture surrounds them. In environments where people are rushed, unheard, or treated as expendable, the technology will inherit that tone. In environments where safety is inspectable, training is real, and accountability is shared, the technology will reinforce trust. These machines will work near people, and their behavior will be read as intent, even when it is not.

A system that can do more than people expect but cannot explain itself, constrain itself, or earn acceptance will stall. A system that does less at first but behaves predictably, reports honestly, and improves in the open will scale.

Where Leadership Emerges

Regional diversity will also shape where leadership emerges. Leadership will not belong only to those who build the most advanced hardware or train the largest models. It will belong to those who create trust at scale. That can happen through safety certification. Through facilities integration. Through workforce training. Through governance frameworks that others rely on. Influence in this field will accrue to those who reduce friction for everyone else.

This creates an opening that is easy to miss. Late movers are not locked out if they choose their leverage points well. Smaller players are not excluded if they show up prepared. The field is still early enough that standards, interfaces, and norms are being set by those who bring evidence rather than ambition. The winners will not be those who claim inevitability. They will be those who make progress legible.

Choice under momentum is the real theme of this moment. Momentum will carry humanoids forward regardless. Labor pressures will persist. Infrastructure demands will rise. Automation will expand. The question is not whether these systems will exist. The question is what kind of presence they will have.

That choice shows up in design decisions that either narrow or widen future options. It shows up in whether systems are built to travel across borders or locked into one context. It shows up in whether logs are treated as liabilities to hide or assets to share. It shows up in whether deployment is framed as replacement or as reallocation.

From an ecosystem perspective, the work is to keep doors open without losing center. To support speed without sacrificing discipline. To allow dual use without letting military framing swallow civilian trust. To welcome capital without letting hype distort timelines. To build infrastructure that carries learning forward instead of resetting it every cycle.

This is a call for precision. Precision in claims. Precision in constraints. Precision in accountability. Precision is what lets a field move fast without breaking trust. It is what allows different regions to cooperate without erasing differences. It is what allows innovation to compound rather than fragment.

The difference was never the technology. It was the posture of the people deploying it.

Humanoids will become part of how work is done, how infrastructure is maintained, and how resilience is built. That outcome is already in motion. What is still open is whether they become quiet contributors that raise safety and capacity, or visible symbols that trigger resistance and heavy-handed control. That line will be drawn by behavior long before it is drawn by law.

The field does not need grand narratives to get this right. It needs consistent practice. Systems that behave within declared bounds. Programs that publish what they learn. Leaders who are willing to say no when a use case undermines long-term trust. Institutions that reward evidence over theater.

If those elements hold, humanoids will not arrive as a shock. They will settle in as part of the background infrastructure that keeps complex societies running. Not perfect. Not invisible. But accepted.

That is the opportunity at this stage. Not to predict the future, but to shape the conditions under which it unfolds. The work ahead is not about proving what humanoids can do. It is about proving how they belong.

The momentum is real. The shape is still a choice.

Epilogue

Humanoids 2.0—What Comes After First Contact with the Real World

I WROTE THIS book to frame the work ahead, not to declare victory. We are leaving the safe part of the curve. The next chapter of humanoids will be defined by what happens when prototypes touch the real world, when design choices carry moral weight, and when scale starts to shift economies.

We should enter this phase with clear eyes and steady hands.

The Real-World Deployment Phase

The lab gives you a signal. The field gives you truth. Real sites are noisy, crowded, and full of edge cases that no test plan can fully capture. Doors stick. Floors are wet. Lighting changes by the hour. People behave in ways that confound tidy assumptions. This is where humanoids either earn trust or lose it.

Deployment exposes gaps quickly. Battery life looks different when a robot climbs stairs, carries weight, and idles in queues. Manipulation looks different when parts are slightly out of spec or when packaging tears. Perception looks different when reflective surfaces bloom, and occlusions stack up. The fix is not magic. It is faster iteration, tighter loops between operators and engineers, and metrics that tie directly to useful work.

The early pilots matter more than any keynote stage. They set public expectations and investor confidence. They also teach founders how to run real operations. Uptime targets need honest accounting. Mean time between failures must include full recovery. Safety cases must hold up under shift pressure, not just in a demo slot.

Success on day one is steady improvement under load. The most credible deployments will show week-over-week gains, shrinking intervention rates, better cycle times, and clearer boundaries on what the robot can and cannot do. Publishing those curves, even when they are not flattering, builds trust.

This phase also forces pragmatic choices. Scope creep is the enemy. Pick tasks with stable interfaces, known failure modes, and measurable value. Start where the environment is structured enough to be learnable but real enough to matter. Hold the bar on safety. Say no to flashy but fragile.

The Moral Decisions and Ethical Frameworks Phase

As autonomy grows, decisions shift from people to systems. Even with a human in the loop, intent, accountability, and acceptable boundaries need to be explicit. This begins at design time, not after an incident.

Define harm hierarchies and safe defaults. Set conservative stop conditions. In practice, this can mean halting if a human enters a defined radius, defaulting to the slowest safe speed in mixed environments, and requiring manual confirmation before actions that involve close physical contact. Prefer de-escalation over task completion when risk rises. Require consent where humans are directly affected. Protect privacy by default. Log decisions with enough fidelity to audit after the fact. Make it possible to explain why an action was taken.

Checklists, playbooks, and tooling should fit into daily workflows. It also means red-team exercises, adversarial simulation, and independent reviews that do not report to the same people who own deployment targets. If ethics only live in policy documents, they will not survive a busy shift.

No single framework will be final. Capability will outpace policy. The companies that bake ethics into design from day one will hold an edge. They will catch problems earlier, avoid brittle shortcuts, and move faster because their systems are built to be inspected and corrected.

We should expect evolving norms. Some rules will become standard across the industry. Others will remain domain-specific, tuned to hospitals, factories, or public spaces. What matters is shared principles with local adaptation—rules that are enforceable, auditable, and tied to consequences.

The Global Transformation Phase

Scale changes the conversation. Once humanoids are integrated into operations at thousands of sites, the effects compound. Output per worker rises. Downtime costs fall. New services become possible in places that could not support them before. At the same time, roles shift, and some communities feel pressure first.

The benefits can be broad, but only if adoption is matched by adaptation. Workforce transition is budgets, programs, and real timelines. Companies that succeed will plan reskilling early, not as public relations after deployment. They will move people from dull, dangerous, and repetitive tasks into maintenance, supervision, quality, and data roles, and they will measure the outcomes.

Capital allocation will change as well. Supply chains for actuators, sensors, batteries, and service tooling will become strategic assets. Nations will define interests around standards, safety, and manufacturing. Countries and companies that align technology with human values will be more resilient. They will export systems people want to work with, not just tolerate.

This phase is less about what the robot can do and more about how we choose to use it. Procurement rules can reward safety records and transparent reporting. Insurance can price real risk and encourage better practices. Regulation will matter. Proactive frameworks that set safety standards and audit mechanisms enable faster, more confident deployment than reactive rules written after failure. The difference is trust.

If gains concentrate in a few hands, backlash will grow. If gains spread, support will grow with them. The leaders in this era will be those who match performance with responsibility, speed with care, and ambition with clarity.

I trust field hours over stage demos. That is how I was trained, and it is how I plan to contribute going forward. The first wave brought us to the start line. Humanoids 2.0 is about showing up in the places that keep the world running and doing useful work, safely, day after day.

We reached the start line. The rest of the story belongs to Humanoids 2.0.

About the Author

Modar Alaoui is the founder and general partner of ALM Ventures, a Silicon Valley venture capital firm focused on humanoids, physical AI, and spatial intelligence. He is also the founder and chair of the Humanoids Summit, a global forum that brings together leading builders, investors, and operators shaping humanoid robotics, with global editions in Silicon Valley, London, Tokyo, Seoul, and a growing international network. He operates at the intersection of capital, strategy, and infrastructure, and has played a central role in advancing the humanoid robotics ecosystem toward commercial deployment and mass production.

Modar is the founder and chair of the Bay Area Robotics Association (BARA), a member-led industry organization established to connect capital, startups, enterprises, and public sector stakeholders across Silicon Valley and global robotics hubs. He is the founder of International Humanoids Day, a global initiative advancing "humanoids for humanity," aligning industry leaders around ethical development, global standards, and international collaboration in humanoid robotics. He contributes to policy efforts shaping US humanoid strategy and long-term deployment frameworks, helping strengthen US competitiveness in physical AI.

Earlier in his career, Modar founded Eyeris, where he developed computer vision and sensor fusion AI systems for in-cabin sensing. His work combined software and hardware across perception and edge compute, providing hands-on experience with real-world systems operating around people.

He studied behavioral science and communication at Concordia University in Montreal, forming the foundation for his work in human behavior understanding and machine perception. This combination of behavioral insight and physical AI systems continues to inform his work in humanoids and real-world autonomy.

His work and perspectives have been featured in major media outlets including Bloomberg, CNBC, *The Wall Street Journal*, Reuters, *Wired*, Scientific American, The New Yorker, and *Forbes*. In *The Humanoid Economy*, he draws on his role in building the humanoid robotics ecosystem to frame the next phase of industrial transformation.

Acknowledgments

I'm grateful to the founders, builders, and researchers operating across humanoids and physical AI. Their work is the foundation of this book. Many have been working toward this moment for years, driven by the belief that intelligence would eventually take physical form—a conviction I have shared alongside them for decades. Their courage, dedication, and years of grit continue to shape the field as it integrates into the fabric of daily lives.

I extend my gratitude to my team, whose pursuit of excellence, execution, and discipline has been instrumental behind the scenes.

I'm thankful to Mikaela and Randy for their guidance and perspective during the development of the manuscript. Their input helped bring structure and clarity to the work while keeping it authentic to my own experience.

Index

Numerics

12/12 symmetry, 85–86, 98

A

acceptance, humanoids
- culture, 175–177
- constraints, 172–175
- exposure reveals, 170–171
- fear of human replacement, 155–159, 168–169
- and human space, 165–167
- human work structure changing, 159–162, 168–169
- presence demands, 167
- reevaluation, 162–165, 169–170

accountability, 171–173
- economic interface and trust, 148
- structures design, 160–161

actuation
- architecture
  - and power, 108–110
  - and safety, 125
- sensors and, 29

AI. *See* artificial intelligence

alignment
- and costs
  - fragmentation, 186–188
- cross-regional, 94–97
- forces
  - AI and hardware, 41–42
  - market conditions, 42–43
  - power of, 43–44

ALM Ventures, xv, 97

anchoring
- cultural, 85–86, 98, 100
- regional, 80
  - hidden inefficiencies, 81–83
  - working layer, 81

anthropomorphism, 116
anxiety, inherited, 156
architecture, 103–104
decisions, 129–130
tradeoffs, 132
failure, default state, 121–124
five pillars of
actuation and power, 108–110
balance and locomotion, 117–119
compute and control, 110–112
manipulation and dexterity, 113–117
sensing, 104–108
generalization, 134
constraint on scale, 137–138
demonstration and deployment, 136
difficulties, 135
optimization, limits of, 135–136
real environments, variability of, 134–135
stable and predictable, 136–137
integration, 119–121
outcome, reliability, 127–130
safety constraint, 124–127
artificial intelligence (AI), xiii, 3, 6, 17
embodied, governs, 51–52, 56
and hardware, 41–42
models
large-scale, 54
perception models, 30–32
physical, capital funding, 140–145
real-world systems, xvi
research, 33
systems, 35–36
Asia, xv, xvi, 70, 94
optimize coordination and execution, 194–195
attention, strategic ecosystem, 183–186
audio models, 4

B

balance, architecture and locomotion, 117–119
BARA. *See* Bay Area Robotics Association
Bay Area Robotics Association (BARA), 80–85, 95, 99
behavior over time, 77–78
bipedal locomotion, 118–119

C

capital funding
and physical AI, 140–141
and risk model, 142–145
capital markets learning patterns, 152–154
Center of Excellence, 84, 98
center of gravity, global movement, 61–62
changing, human work structure, 159–162
China, 70, 93

cloud systems, 112
co-evolution, humans and technology, 22–23
  loop in practice scale, 23–26
collision, physical AI and capital funding, 140–141
commitment, fund, 47
companies stall, humanoid, 150
  metrics, mislead, 151
  survive, 152
complementary regional strengths, 93–94
compute, architecture and control, 110–112
constraint
  architectural
    generalization, 134–138
    safety, 124–127
  culture, 172–177
construction, portfolio, 51
contact-rich environments, 108
control
  architecture and compute, 110–112
  demonstrations, 127–128
  environments, 118
  systems, 55, 112, 125, 131, 135
convergence, global, 38–39
coordination
  alignment and costs, 186–188
  human responsibility, 163–165
cost, ecosystem, 186, 187
cross-regional alignment, 94–97
cross-region investing model, 58
cultural acceptance, humanoids, 175–177
  constraints, 172–175
  enters human space, 165–167
  exposure reveals, 170–171
  fear of human replacement, 155–159, 168–169
  hesitation, misreading, 175
  human work structure changing, 159–162, 168–169
  presence demands, 167
  reevaluation, 162–165, 169–170
cultural anchor, 85–86, 98, 100
cultural environment, 43
cumulative effect, 45

D
decisions
  act, 46–48
  architectural, 129–130
    tradeoffs, 132
  moral, 200–201
deep learning perception, 54–55
degradation, sensing, 106
demonstrations, 34
  architecture and deployment, 136
  controlled, 127–128
  large-scale, 39, 41–42
deployment
  approaches real-world scale, 80
  architecture and demonstration, 136
  governance, 78–80

deployment (*continued*)
humanoid, 159–160
phase, real-world, 199–200
structure, 75–76
missing layer in, 77–78
dexterity, 4
architecture and manipulation, 113–117
difficulties, generalization, 135
diversity, regional, 191
leadership emerges, 195–197

E

early humanoid systems, 111
economy, humanoid
architecture. *See* architecture
companies stall, 150–151
cultural acceptance. *See* cultural acceptance, humanoids
designing for humanity, 7–8
economics point, 8–10
ecosystem. *See* ecosystem, absence of
funding. *See* funding, human-like machines scale
global stakes. *See* global stakes
human acceptance, 14–15
forming, 15–16
roles, 17–22
and scale, 16–17
infrastructure. *See* infrastructure, movement
loops, humans and technology, 22–26
making fit, 10–11
robotics. *See* robotics
senses, 11–14
ecosystem
gaps revealed, 60–61
global model, 68–70, 73, 93
infrastructure, 59–60
strategic attention, 183–186
ecosystem, absence of, 27
aligned forces
AI and hardware, 41–42
market conditions, 42–43
power of alignment, 43–44
global convergence, 38–39
humanoids momentum
isolated breakthroughs, 28–29
progress and readiness, 29–30
path to act
build commitment, 47–48
decision to act, 46–47
difficulty to inevitability, 45–46
scattered efforts without a system, 30–31
cost of independence, 31–33
signals, misread, 33–37
talent without connection, 39–41
electromechanical and mechatronic tools, 3
embodied AI, xiv, xvi
governs, 51–52, 56, 60, 90
recognition, 46–47
embodied intelligence, xiv, xvi, xvii, 28, 33, 37, 38, 40, 46, 47, 72, 75, 88, 179

energy efficiency, 109
environments
contact-rich, 108
controlled and human, 118
cultural, 43
humanoids, 11–14
robotics, 3–7
sensing, 12
variability of real, 134–135
ethical frameworks
phase, 200–201
Europe, xv, xvi, 69–70, 93
optimization, 194–195
policy framework, 96
evidence, strategy, 184–184
expensive, human labor, 161

F

failures
humanoid, 105–106
integration, 120–124
visible, 114
fear of human
replacement, 155–159
feedback loops,
integrated, 113–115
first edition, Humanoids
Summit, 65–68
fits, making humanoids, 10–11
five pillars, architecture
actuation and power, 108–110
balance and
locomotion, 117–119
compute and control, 110–112
manipulation and
dexterity, 113–117
sensing, 104–108
forces alignment, AI and
hardware, 41–42
foundation models, 3, 41
founders
assumptions, 60–61
ecosystem. *See* ecosystem,
absence of
funding
economic
interface, 145–150
humanoid companies
stall, 150–151
physical AI, 141–145
infrastructure. *See*
infrastructure, movement
fragility, human-scale, 180
fragmentation, 31–32, 95
alignment and costs, 186–188
perception gap, 34
risk of, 183
technical, 39
framework
HOMES, 78–80
phase, ethical, 200–201
policy, 90–92
humanoids fit human
environments, 91–92
standards in practice, 92
framing, practical, 87–88
full-body
control integration, 30–31
coordination, 4
movement analysis, xiv
fund formation, infrastructure,
59
early-stage robotics, 55, 59
geography, 52–53

funding, human-like machines scale, 139–140
capital markets learning patterns, 152–153
economic interface, 145–147
accountability and trust, 148
progress, shapes, 148–150
humanoid companies stall, 150–151
physical AI
capitals, 141–142
different risk model, 142–145

G

gaps revealed, ecosystem, 59–61
generalization, architecture, 134
constraint on scale, 137–138
demonstration and deployment, 136
difficulties, 135
optimization, limits of, 135–136
real environments, variability of, 134–135
stable and predictable, 136–137
trust, 136–137
general-purpose design
humanoid, 131–132
platforms, 5–6
systems, 132–133
geography
fund, 52–53
process, 96
global
convergence, 38–39
ecosystem, 68–70, 73, 93
movement, center gravity, 61–62
robotics ecosystem, 71
Summit platform, 70–75
transformation phase, 201–202
Global Expansion Program, 71–72
global stakes, 180–181
coordination or fragmentation alignment and costs, 186–188
ecosystem, standards and costs, 183–186
governance, infrastructure, 191–194
leadership emerges, 195–197
operational, responsibility, 195
regional gravity, 188–189
strategic national
assets, 180–181
attention, 181–182
fragmentation, risk, 183
governance
deployment, 78–80
infrastructure, 191–194
models, 90
gravity
global movement, 61–62
regional patterns, 188–190

H

hardware, AI and development, 34, 41–42
hidden inefficiencies, 81–83

HOMES. *See* Humanoid Operating Model for Environment Safety
human
acceptance, 14–15
forming, 15–16
roles of, 17–22
and scale, 16–17
co-evolution, 22–26
environments, 118
and humanoids
economic interface, 145–150
fits, 91–92
standards, 8–11
labor, expensive, 161
organizations, 148
replacement, 155–159
space, humanoid enters, 166
fear changes form, 168–169
presence demands, 167
work
responsibility, 163–165
structure, changing, 159–162
humanity, designing, 7–8
human-like machines, funding scale, 139–140
capital markets learning patterns, 152–153
economic interface, 145–147
accountability and trust, 148
progress, shapes, 148–150
humanoid companies stall, 150–151
physical AI
capitals, 141–142
different risk model, 142–145
humanoid, economy
architecture. *See* architecture
companies stall, 150–151
cultural acceptance. *See* cultural acceptance, humanoids
designing for humanity, 7–8
economics point, 8–10
ecosystem. *See* ecosystem, absence of
funding. *See* funding, human-like machines scale
global stakes. *See* global stakes
human acceptance, 14–15
forming, 15–16
roles, 17–22
and scale, 16–17
infrastructure. *See* infrastructure, movement
loops, humans and technology, 22–26
making fit, 10–11
robotics. *See* robotics
senses, 11–14
Humanoid Operating Model for Environment Safety (HOMES), 78–80
humanoids 2.0, 199–202
Humanoids Act proposal, 90–92, 98
Humanoids Summit, xv, 61–62, 85, 94
establishing, 62–63

Humanoids Summit (*continued*)
first edition, 65–68
global ecosystem and platform, 68–75
Humanoids Summit Global Expansion Program, 71
human-scale fragility, 180

I

industrial robots
isolation, 104
thrive, 6
traditional, 125
inefficiencies, hidden, 81–83
inevitability, difficulties, 45–46
infrastructure, movement, 49–50
cultural anchor, 85–86
12/12 symmetry, 86
shift in perception, 86–90
ecosystem, gaps revealed, 59–61
governance, 191–194
Humanoids Summit, 61–62
establishing, 62–63
first edition, 65–68
global ecosystem engine, 68–70
global summit and platform, 70–75
policy framework, 90–91
humanoids fit human environments, 91–92
standards in practice, 92
regional anchoring, 80
BARA, 83–85
hidden inefficiencies, 81–83
working layer, 81
regional nodes, ecosystem
complementary regional strengths, 93–94
cross-regional alignment, 94–97
strategy in practice, 97–102
structure beneath
deployment, 75–76
governance, 78–80
missing layer in, 76–78
Venture Capital Fund, 50–51
investment thesis, 51–53
path and fund, 53–54
robotics struggles, 54–59
inherited anxiety, 156
insurance and liability, 185
integration
architecture, 119–121
feedback loops, 113–115
full-body control, 30–31
intelligence
adaptable, 55–56
spatial, 51–52, 56
interface, economic humans and humanoids, 145–152
International Humanoids Day, xv, 85–90, 98
International Women's Day, 88
investment
cross-region, 58
strategies, 55–56, 58
thesis, 51–53
isolation
breakthroughs, 28–29
industrial robots, 104

J
Japan, 93
judgment, human
responsibility, 163–165

L
language models, 4
large-scale AI models, 39,
41–42, 54, 55
Latin America, 71
launching
Venture Capital Fund,
50–51
path and fund, 53–54
robotics struggles, 54–59
leadership, regional
diversity, 195–197
learning system patterns, 55
capital markets, 152–154
robotics, 3–5
liability
governance, 192–193
insurance and, 185
locomotion, 4, 8
architecture, balance,
117–119
breakthrough in, 28–29
systems, 35
logistics
hub, 86–88
robot, 3, 35
London, 70–71, 94, 98
loops, controlled demonstrations
humans and technology
in practice, 23–24
shapes scale, 24–26
integrated feedback, 113–114

M
machine
learning models, 33–36
physical language of, 17–22
manipulation
architecture and
dexterity, 113–117
systems, 35
market
capital learning
patterns, 152–154
conditions, 42–43
recalibrate, 144–145
measurement, economic
interface, 147
media amplification, 156
metrics, companies, 151
misaligned perspectives, 78
momentum, before the
humanoid
isolated breakthroughs,
28–29
progress and readiness
gap, 29–30
moral
decisions, 200–201
operational responsibility,
195
movement, infrastructure
cultural anchor, 85–86
12/12 symmetry, 86
shift in perception, 86–90
ecosystem, gaps
revealed, 59–61
governance, 191–194
Humanoids Summit, 61–62
establishing, 62–63

movement, infrastructure (*continued*)
first edition, 65–68
global ecosystem engine, 68–70
global summit and platform, 70–75
policy framework, 90–91
humanoids fit human environments, 91–92
standards in practice, 92
regional anchoring, 80
BARA, 83–85
hidden inefficiencies, 81–83
working layer, 81
regional nodes, ecosystem
complementary regional strengths, 93–94
cross-regional alignment, 94–97
strategy in practice, 97–102
structure beneath
deployment, 75–76
governance, 78–80
missing layer in, 76–78
Venture Capital Fund, 50–51
investment thesis, 51–53
path and fund, 53–54
robotics struggles, 54–59

N

national assets strategy, 180–183
national stereotypes, 188–191
negotiation, persistence of, 175–176
nodes, regional, 93

O

operational responsibility, 195
optimization, limits of, 135–136
orchestration, power, 110
organizational behavior interfaces, 148–149

P

pace robotics, 4
path, recognition, 88–89
patterns, capital markets learning, 152–154
payoffs, 8–10
perception
AI models, 30–32, 121
changes, integration, 122
deep learning, 54–55
gap, fragmentation, 34
public, 176
sensing, 105–106
systems, 31–32, 35, 55, 135–136
perspectives, misaligned, 78
physical AI
capital funding, 140–141
language of machines, 17–22
risk compounds and models, 141–145
policy framework, 90–92
portfolio construction, 51
power
actuation and architecture, 108–110
of alignment, 43–44
management reinforce, 110
orchestration, 110
practical framing, 87–88

practice
  standards in, 92
  strategy, infrastructure, 97–102
predictability system, 136–137
presence, human, 167–169
pricing interface models, 148–150
progress
  capital mismatch, 143–144
  interface, pricing models, 148–150
  technical, 58
public perception, 176

R

real-world AI systems, xvi
  constraint, generalization, 134
  deployment phase, 77, 199–200
  environments, variability of, 134–135
recalibrate, markets, 144–145
recognition
  embodied intelligence, 46–47
  path, 88–89
recovery, design of, 129
reevaluation, humanoids, 162–165, 169–170
regional anchoring, 80
  hidden inefficiencies, 81–83
  working layer, 81
regional diversity, 191
  leadership emerges, 195–197
regional gravity patterns, 188–190
regional nodes, 93
  complementary regional strengths, 93–94
  cross-regional alignment, 94–97
regional strengths, complementary, 93–94
reinforcement learning methods, 33, 36, 41
reliability
  architectural outcome, 127–130
  and safety, 131–132
replacement, human fear, 155–159
research-driven systems, 58
responsibility
  human work, 163–165
  operational, 195
risks
  capital mismatch progress, 144–145
  compounds, physical, 141–142
  fragmentation, 183
  models, physical AI, 142–145
robotics, 2–3
  and environments, 6–7
  general-purpose platforms, 5–6
  humanoid. *See* humanoid, economy
  learning, 3–5
  movement, 5
  rules and predictable routines, 2–3
  struggles, 54–59
  traditional industrial, 125

S
safety
constraint,
architectural, 124–127
deployment and standards,
75–76, 91–92
reliability and, 131–132
scale, humanoids
funding, 139–140
capital markets learning
patterns, 152–153
economic interface, 145–147
accountability and
trust, 148
progress, shapes, 148–150
generalization, 137–138
humanoid companies
stall, 150–151
physical AI
capitals, 141–142
different risk
model, 142–145
scattered efforts, 30–31
cost of independence, 31–33
sensing
architecture, 104–108
humanoid, 11–14
sensors, 34–35
and actuators, 29
fusion, 12–14
and hardware
improvements, 36
humanoids, 11–12
signals
capital markets, 152–154
humanoids misread, 33–37
Silicon Valley, xv, xvi, 46, 50, 53,
68, 70, 94, 95
first edition, 65, 68
mindset, 28
software risk, 141–144
South Korea, 70, 71, 93
spatial AI, 3
spatial intelligence,
51–52, 56, 60
specificity, humanoid
design, 131–134
stability system, 136–137
stall, companies
humanoid, 150–152
standards
deployment
behavior over time, 77–78
misaligned perspectives,
78
and safety, 75–76, 91–92
in practice, 92
strategy, 184–185
startups, platforms, 30, 39
strategy
national assets, 180–181
attention, 181–182
fragmentation risks, 183
practice,
infrastructure, 97–102
structure
beneath deployment, 75–76
technical progress, 29–30
struggles, robotics, 54–59
Summit
first edition, 66–67
global platform, 70–75
Humanoid
first edition, 65–68
global ecosystem and
platform, 68–75

Humanoids, 61–62
establishing, 62–63
scaling, 68–70
systems, humanoids, 119–121

T
task-specific
architectures, 132
humanoids, 131–132
systems, 132–133
technology
co-evolution of, 22–26
foundations, 41–42
fragmentation, 39
progress, 58
and structural readiness, 29–30
revolutions, 36
thesis, investment, 51–53
three-region model, 70
time, compute and control, 110–112
Tokyo, xvi, 70–71, 94, 98
tradeoffs, 119
architectural decisions, 132
humanoid development, 44–45
tradition, industrial robotics model, 11, 24, 125
transformation phase, 201–202
transformative technologies, 37
transition, deployment and demonstration, 136
transmission systems, 42
trust
economic interface and accountability, 148
inspectability, 184–185
over time, system, 87, 136–137
turning point robotics, 2–6
two-handed manipulation compounds, 115–116

U
United Nations, 88
United States, xv, xvi, 93, 94
deployment strategies, 96
ecosystem, 68–69
global convergence, 38–39
optimize speed and capital formation, 194
talent without connection, 39–41
US Center of Excellence, 92
US Humanoids Act, xv, 90

V
variability, real environments, 134–135
venture capital community, 29, 34–35, 43
Venture Capital Fund, 50–51
investment thesis, 51–53
path and fund, 53–54
robotics struggles, 54–59
visible attention and failure, 113, 114
visual models, 4

W
workforce implications, 185
working layer, 81
World Environment Day, 88
world models, 3
World Space Day, 88